PRINTING
TECHNOLOGY

印刷工艺

深圳市艺力文化发展有限公司 编

Print the World, Read the Heart

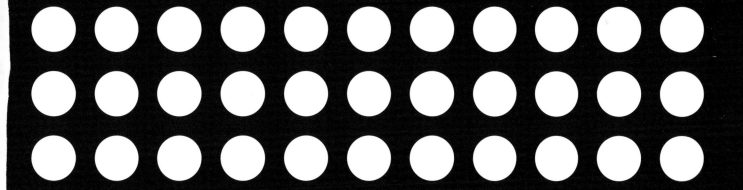

华南理工大学出版社
SOUTH CHINA UNIVERSITY OF TECHNOLOGY PRESS

·广州·

图书在版编目（CIP）数据

印刷工艺 = Printing technology：英文 / 深圳市艺力文化发展有限公司编. — 广州：华南理工大学出版社，2014.3
ISBN 978-7-5623-4177-2

Ⅰ.①印… Ⅱ.①深… Ⅲ.①印刷－生产工艺－英文 Ⅳ.①TS805

中国版本图书馆CIP数据核字（2014）第041317号

印刷工艺　Printing Technology
深圳市艺力文化发展有限公司　编

出 版 人：	韩中伟
出版发行：	华南理工大学出版社
	（广州五山华南理工大学17号楼，邮编510640）
	http://www.scutpress.com.cn　　E-mail: scutc13@scut.edu.cn
	营销部电话：020-87113487　87111048（传真）
策划编辑：	赖淑华
责任编辑：	陈　昊　孟宪忠
印 刷 者：	深圳市汇亿丰印刷包装有限公司
开　　本：	889mm×1194mm　1/16　印张：22.25
成品尺寸：	215mm×285mm
版　　次：	2014年3月第1版　2014年3月第1次印刷
定　　价：	320.00元

版权所有　盗版必究　　印装差错　负责调换

PREFACE

Communication is mutating. When we talk about communication we are now doing it using all the facets of this word: whether they are written, visual, sensory or interactive. Communication floods everything and we are now in an environment where access to production techniques is within reach of anyone.

From the point of view of our profession, now is the time of ideas; of researching new ways of expression to make sure they reach the recipient in a way that is most clear and famous — either by using the latest technology or reclaiming forgotten artisan techniques.

I started my career studying the processes and production techniques in the graphic arts. These studies gave me access to a world of possibilities to communicate ideas: the systems of graphic production. In a moment in time where communication through virtual environments is so powerful and can reach everywhere, personal communication with a physical presence obtains a special intensity, highlighting every detail and every decision we make.

This is one of the reasons why having technical knowledge and curiosity to reach new solutions is crucial. It is necessary to create memorable products, and those memories are built by notable facts. Those exceptional facts stick in our minds, and it is for this reason that the elements of surprise and innovation are essential in the creative work. The production process helps us reach these elements of surprise and innovation.

Working with our senses is a way to create a memorable experience, and doing so is a success for whatever product we are building. Both technique and technology work in our favour by using specific technologies in the design of the project, as well as using artisan techniques in visually innovative applications. They are tools we must learn to use to express our ideas.

Communication starts with a need, and this is where we begin our work as designers. Our clients often need to distinguish themselves from their competitors. But in a mature environment, simply mentioning that you are different is no longer enough: you need to show it and that first contact is critical to do it. An all round creative strategy, an effective design and a good selection in the production process are essential for optimal results. This selection must be responsive to the client, to the budget and also to the environment. More often than not clients have a limited budget, but great results can be achieved with appropriate research, a powerful design fitting the objectives, and a good selection of materials and production processes. We live in a world with limited resources and unnecessary spending makes no sense. Designing responsibly that we can get good results, and this book has plenty of examples.

We reach the best solutions working as a team with our printers. Printers are the specialists in their field and thus they become our own professional consultants. We must see them as someone to reach out for help, someone who will help grow the value of our design. They are a key element to a job well done, and we must treat them with confidence and professional respect.

Before I finish I would like to give an example by our studio which was published in this book. It is the project for the gastro-bar Betlem, established in a centennial location. Our choice of materials was influenced by the architecture of the space and the materials used in the buildings of the restaurants; these defined very naturally which options we could choose. The restaurant had a bar made of marble and metal, and they also kept their older wooden furniture and decorative elements. We wanted to translate these visually with the materials and printing techniques used. This way we created consistency in the communication, and it established a direct link between design and physical space. With the business card you can take a bit of the experience in the restaurant with you.

We must explore, research and try to improve on visual and sensory elements at all times, but never forget that design is much more than just form; it is a good idea and we must know how to execute it.

Ferran Mitjans

Partner at Toormix, design studio based in Barcelona

CONTENTS

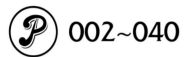 002~040

- 002　Concrete Business Cards
- 004　Nördik Impakt — Global Communication
- 007　T & J Wedding Invitation
- 010　Motu
- 011　Thermo Business Cards
- 014　Murmure Identity 1
- 018　Steve Shine Brand Identity
- 020　The Bold & Brave
- 024　Popular Front Business Cards and Stationery
- 025　Zelo and Bacio Menus
- 026　Tuaca Brand Kit
- 028　Andaz 1901
- 030　Infamous Media Kit
- 031　Pink PSP Media Kit
- 032　PS3 / PSP Media Kit
- 034　Identity for Treverket
- 036　SIZ Bottle Box
- 038　Deck
- 040　Kew Breathing Planet

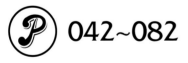 042~082

- 042　Catalogue
- 043　DMWORKOOM Business Card
- 044　Folder
- 046　First Press Palette
- 048　Invitation
- 049　Header
- 050　Second Press Palette
- 052　A Wadded Dream
- 058　Company Profile of Tipografia San Martino
- 062　Pointpath Studios Business Cards
- 064　Black Umbrella Stationery
- 066　Mockbirth EP
- 070　Fa!rbar
- 072　Eyes on Hands on
- 074　MachineDear
- 076　Kitashinchi Bar
- 078　EO Invites
- 081　Paper Bound Engravings
- 082　Wooden Portfolio Box

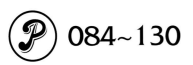

084	La Canya
087	Calendar
088	Kim & Adrian Wedding Suite
090	Come Closer Club
094	Harri Koskinen
098	Meier Seefeld
100	Mangoola
102	Graphic Identity
105	Arkigram — Brand Identity
108	Las Caglias
111	Schuck Juwelier — Brand Identity
114	The Opening of Jacob Jensen Design I DeTao Shanghai
116	Fairy Tale Collection
118	Groovewear Artist Box
120	Wedding Patricia & Gian
122	M-IDEA FOREVER
125	Tomorrow
129	Images Festival 25
130	Sabores Da Vida Gifts

132	La Forma Saporita
139	365 Poster
140	Fairy Tale Themed Wedding Invitation
142	Kinetic Gear Invitation
145	Pop-up Fundraiser Invitation
146	All You Need to Know about Graffiti Is in This Book.
150	Erasable Poster
152	New York-Limited Edition
153	Cleopatra Wedding Stationery Suite
154	Eleanor Wedding Stationery Suite
155	Marianne Wedding Stationery Suite
156	filthymedia Corporate ID
158	Firebelly Holiday Gift
161	10 Jahre Suite212
164	El Palauet Living Barcelona / Corporate Brochure
166	Bottega Veneta
167	Matali Grasset
168	Noguera & Vintró
169	Rough Luxe Hotel

170~208

170	5mm Kahoku Novelty
171	The Silver Ratio Poster
172	Dots & Lines Poster
176	Kushiage Restaurant "Isa" VI
178	Bookbinding
180	Kuli
182	Layers
184	Louis Vuitton Invitation Origami
186	Packing Hacking Series
190	Cliche Christmas Card
192	Culinary Letterpress Posters
195	Greeting Card — MAKE LOVE NOT WORK
197	Invitation — HUNTING LICENSE
198	Typographic Wank
200	Avalanche Print
202	Das Seine. Forschungs Projekt
204	Oh No, Not Sex and Death Again!
206	Julien Hauchecorne Business Card	Diffraction Version
208	Julien Hauchecorne Business Card	Lime-green Version

210~255

210	Mathias Tanguy Brand Visual Identity
213	Levi's, Button Fly
214	N. Daniels Wien
218	Andrew & Kylie — Fiji Wedding Invitation
219	Fiji Wedding Invitation
221	Blustin Design Stationery
224	Playlab Identity
228	Long Play for TWBA
231	Icosahedrons
233	2012 Cheers Coasters
236	Cavalli Corporate Identity
240	Company Brochure "Entrance"
242	E. Company Brochure 2012 Foil Book
246	Green Sky Media Business Card
247	Hear Agency Business Card
249	IDEEËN Autumn Winter Invitation
250	Bodoni Girl
254	The Future is in the Past
255	RockShox Catalogue

 256~288

256	Walking Calendar
258	UAE's Bid Book
260	Personalized Notebook + New Year Card
262	Screenprinted Handbook
264	Autopia Driver Guide
267	American Odysseys: Writings by New Americans
268	Vilcek Holiday Card
269	You & Me Creative's Business Card
270	Pearls & Perfumes
272	Altadonna
274	Coevo
275	Borghetto
276	Earth's Axis
280	Planning on Paper Ring Notebook
281	MediaBite
283	Alejandro & Esperanza Wedding Invitation
284	Esther & Kiko Wedding Invitation
286	Holiday Postcard Business Card
288	White Rabbit Saloon Press Kit

 289~328

289	Suntory Prize for Social Sciences and Humanities
290	Farstad Shipping — 50th Anniversary Book
292	Grøvik Verk — 50th Anniversary Book
294	Jacu Coffee Roastery, Corporate Identity
302	Catastrophic World
304	Special Cards for Special Occasions.
306	Croatian Post Progress Report
308	Graz Opera
312	New Frontier Group
314	Beautiful
316	Long Mark
317	Memory of Jiangnan
318	Betlem Gastro Bar
322	Grafiko
323	Contour
324	Toormix Stationery
326	Emotional Typography
328	A Bazaar Fixx at QT

WELCOME TO
THE PRINTING WORLD

Playing with the notion of scales, Murmure created a set of business cards made of concrete. This material, so characteristic of the environment, was enhanced by using the smallest and most refined communication support. The refinement and the technique required for the typography highlight the harshness and the roughness of the used material.

Concrete Business Cards

Design Agency: Murmure

Client: Self Promotion

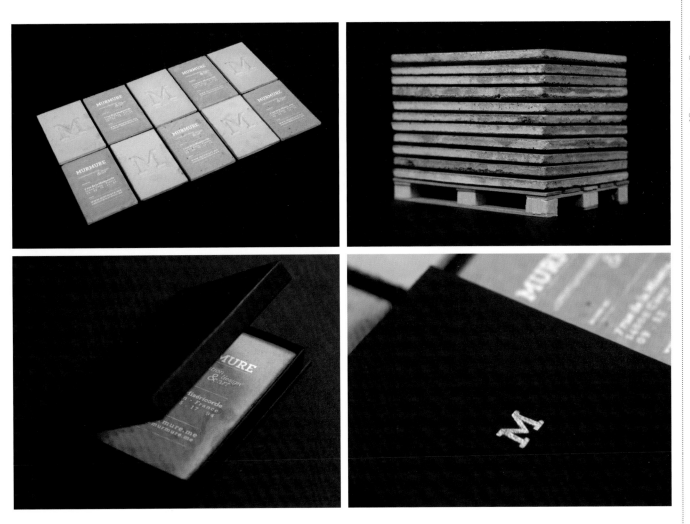

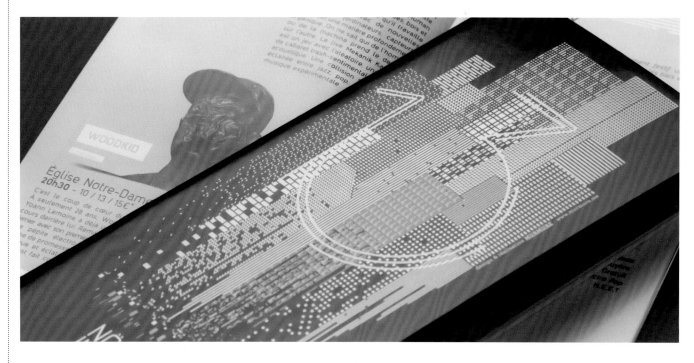

Considered for the last 13 years as the biggest electronic music festival in France, Nördik Impakt wanted to break off visually from its' former edition. Murmure fulfilled the mission of improving the festival's identity, thanks to the creation of a more modern, elegant and conceptual graphic design. Following the theme of a "parallel city", Murmure established a graphic/electronic base underlined by elegant printing processes. Given a fair amount of creative freedom, the agency developed a website in perfect graphic and technological harmony with the festival.

Nördik Impakt — Global Communication

Design Agency: Murmure

Client: ArtsAttack!

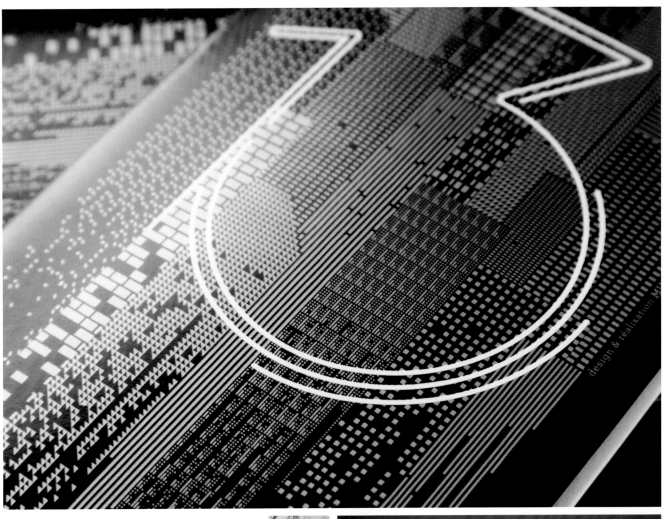

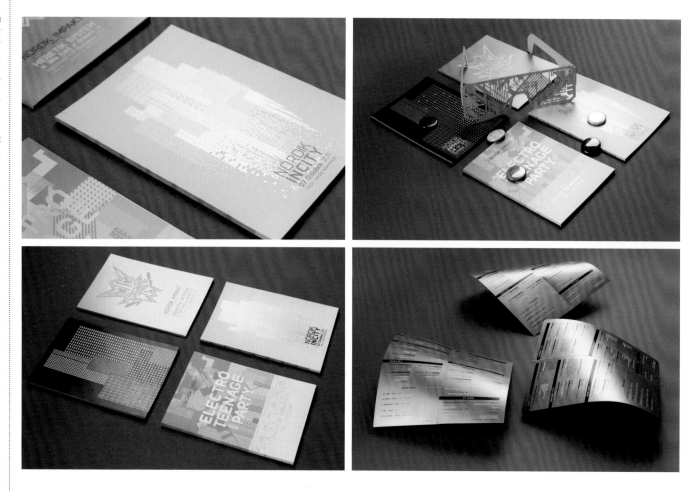

Inksurge was commissioned to design a simple wedding invitation inspired by the banyan tree which was the main idea of the client. Print process — 6 x 6 letterpress finish (front) with 2-color flat design in Bianco White (250 gsm) and flat printing for the back-side (set in Bianco White 130 gsm). Along with that, the designers also silk screened a tote bag in canvas as give-away for the guests.

T & J Wedding Invitation

Design Agency: Inksurge

Designer: Joyce Tai / Rex Advincula

Client: Teresa Liwanag

Printing technology: 6 × 6 letterpress finish (front) with 2-color flat design in Bianco White (250 gsm) and flat printing for the back-side (set in Bianco White 130 gsm) + silk screened tote bag

Printing technology: 6 x 6 letterpress finish (front) with 2-color flat design in Bianco White (250 gsm) + silk screened tote bag and flat printing for the back-side (set in Bianco White 130 gsm)

Motu, a company in New York specialized in public relations, asked the agency to make thermo-sensitive business cards. Experimenting with different printing techniques and conceptual packaging, Murmure created a high-market product.

Motu

Design Agency: Murmure

Client: Motu

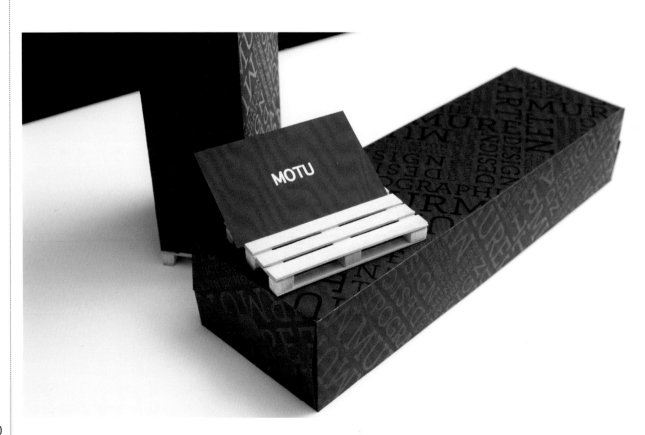

Murmure, a creative agency, developed in 2010 the first thermo-sensitive business cards with the "contact 1" project. In 2011, the contact 2.0 project was the perfect technological and technical culmination, revealing a high-quality commercial potential.

Thermo Business Cards

Design Agency: Murmure

Client: Self Promotion

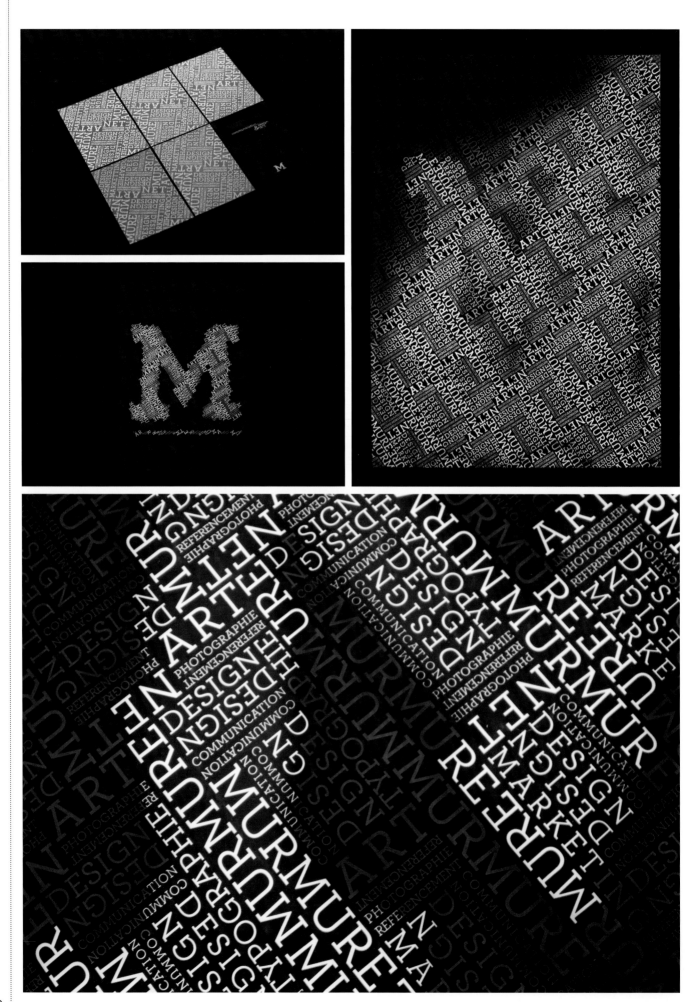

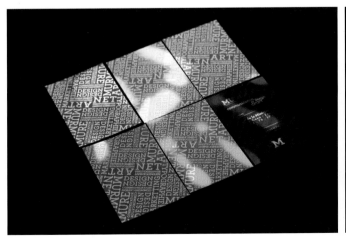

Printing technology: silkscreen

013

Murmure develops an elegant, original and conceptual graphic identity, by using supports and printing procedures which stimulate the senses. Whilst working on technological and conceptual innovation, typographic elegance, the sensuality which emanates from paper supports and textures, the agency reveals what creative and contemporary graphic design can make possible.

Murmure Identity 1

Design Agency: Murmure

Client: Self Promotion

Printing technology: silkscreen

015

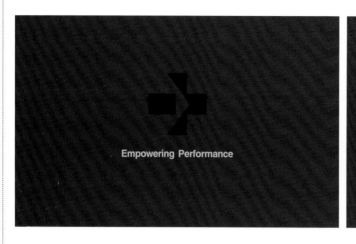
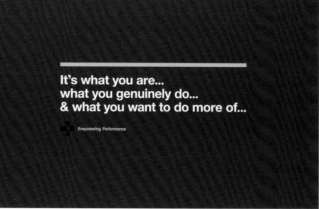

Steve Shine Brand Identity

Design Agency: Analogue

Designer: Mike Johns / Tez Humphreys

Client: Steve Shine

Analogue was approached by Steve Shine with a view to helping articulate his "personal brand proposition" and created a brand look and feel to support his new business aspirations.

The key to the success of Steve's personal brand is being true to whom he is and his known characteristics. Through several interviews and desk research the designers netted down Steve's brand essence to "Empowering Performance".

Steve's business cards are one of his key communication tools. The exterior has an elegant and refined appearance, yet it's vivid and electric blue core serves to echo his powerful and positive personality.

Printing technology: the cards where foil blocked on treble bonded colour plan from GF Smith

A unique collection of 5 different port varieties, created in collaboration with the designers' friends at Bell River Estate.

The design itself was inspired by the rich history of the neighboring wharves, where U.S. sailors on shore lived in Potts Points and Kings Cross helped establish a long surviving culture of working hard and playing harder. A good old-fashioned hole punch became the simple and ingenious solution to bring hand-crafted exclusivity to each individually numbered bottle. The beautifully letterpressed labels are then held in place with a thick industrial rubber band. Illustrator Jeremy Lord, the designers' friends at "The Distillery" letterpress and Clear Image Labels helped in this commercial run. The Bold & Brave Port Co. is a testament of the designers' town. It is never a dull moment, always an opportunity. A place where fortune favors the brave...

The Bold & Brave

Design Agency: Bold-inc

Creative Director: Jon Clark

Design Director: Jon Clark, Jarrod Robertson

Designer: Holly Doran

Printing technology: screen print, offset with a thermographic coating, tear-off

Printing technology: screen print, offset with a thermographic coating, tear-off

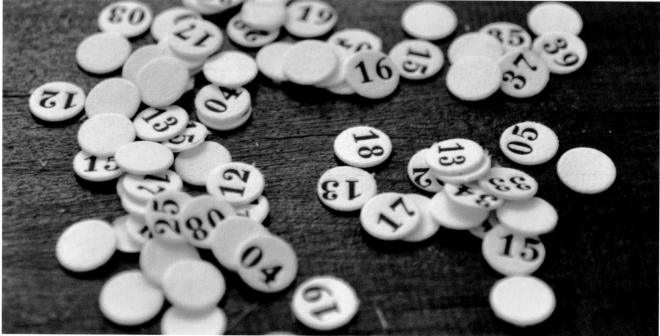

A successful interactive agency, Popular Front was growing its offering to become a holistically-focused branding agency. The new identity is a bold departure, with a logotype and graphic language that is intelligent and dynamic. The wordmark has the feeling of confident precision, with breaks in the geometric letterforms that feel active and in motion. From laser cut business cards to spot varnished, back-printed letterhead, the stationery makes a bold statement to prospective employees and clients alike.

Popular Front Business Cards and Stationery

Design Agency: Cue

Designer: Nathan Hinz and Alan Colvin

Client: Popular Front

For Break Bread Hospitality restaurants, the designers helped to refresh the dining experience by packaging new menu offerings in fabric-covered, hardbound covers. Whimsical illustration elements drawn from interior motifs were used to create covers for Zelo and Bacio menus—to compliment the quality food and drink offering, and add a bit of fun.

Zelo and Bacio Menus

Design Agency: Cue

Designer: Alan Colvin

Client: Break Bread Hospitality

To be relevant, brands need to get noticed, speak in a consistent voice and engage in interesting ways. More and more, brands are activated at the local level. To provide tools and communicate with the field, the Tuaca brand activation guide provides branding elements and programs, packaged in a cool, Tuaca way.

Tuaca Brand Kit

Design Agency: Cue

Designer: Kate Arends and Alan Colvin

Client: Brown-Forman

Printing technology: hard cover notebook wrapped in offset printed silver ink and with a silver foil stamp, laminating, cutting and folding, embossing and debossing

This project saw Fluid produce branding for Hyatt's first "progressive innovative restaurant" in Europe. It was initially released and situated in London; the concept however was later rolled out globally across the chain of The Hyatt Group. The highlight of the project is the range of menus which featured a host of high end print and finishing techniques from embossing to foil stamping.

Andaz 1901

Design Agency: Fluid

Client: The Hyatt Group

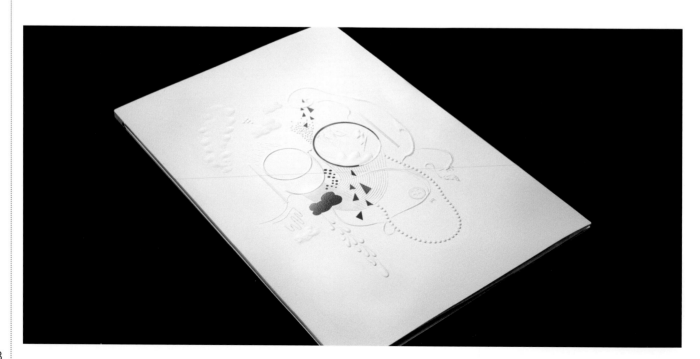

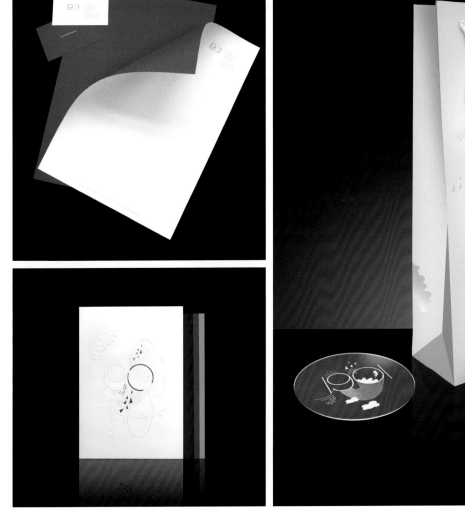

One of Fluid's long-standing clients, SCEE, set the task of producing a media kit for an up and coming new game — Infamous. The kit took the form of a high quality, tactile 24-page publication. The basis of which was to be founded upon the games key themes, comic books, superpowers and electricity. The cover was printed 4 colour process and white on a mirrored lenticular board. The sleeve was blue rubber translucent with a deboss of the logo.

Infamous Media Kit

Design Agency: Fluid

Client: Sony Playstation

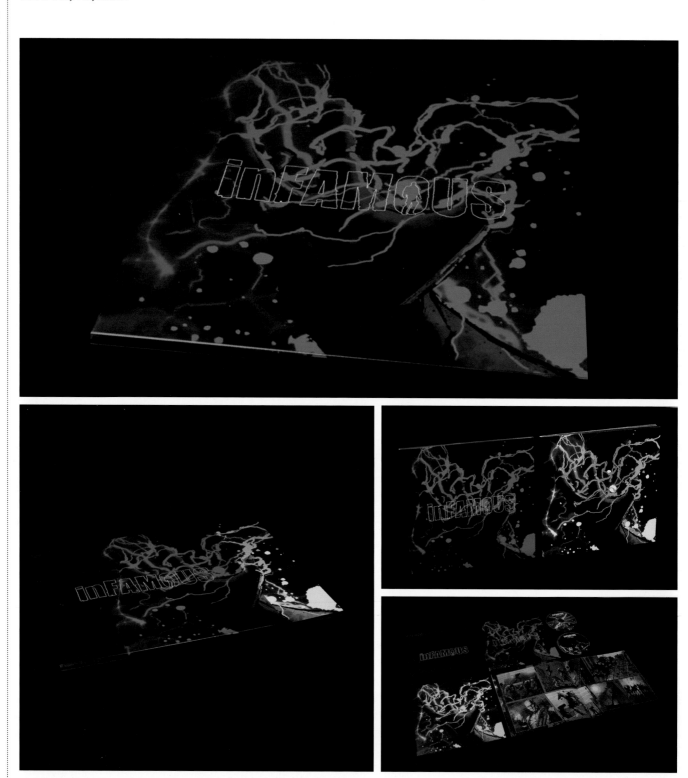

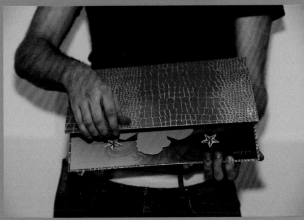

One of Fluid's long-standing clients, SCEE, requested a press kit to be produced for the pink PSP and Pink (the female pop artist). The kit contained a DVD with content and pictures of the PSP and video interviews with Pink herself. The focus for fluid however was the presentation and design of the press kit itself which took the body of a foldout pop up A3 pack alongside spot UV and embossed printing.

Pink PSP Media Kit

Design Agency: Fluid

Client: Sony Playstation

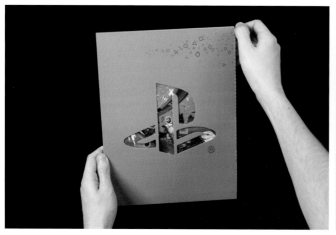
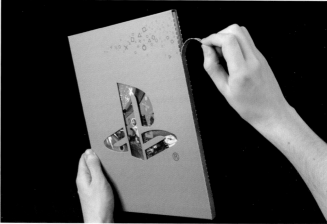

One of Fluid's long-standing clients, SCEE tasked them with the exciting job of coming up with a creative, high quality, vibrant and engaging media kit that packaged together all Playstation 3, PSP and Playstation Network related content into one pack. The end result came in the form of a 68-page oversized A4 book, including an outer diecut slipcase and special print effects throughout.

PS3 / PSP Media Kit

Design Agency: Fluid

Client: Sony Playstation

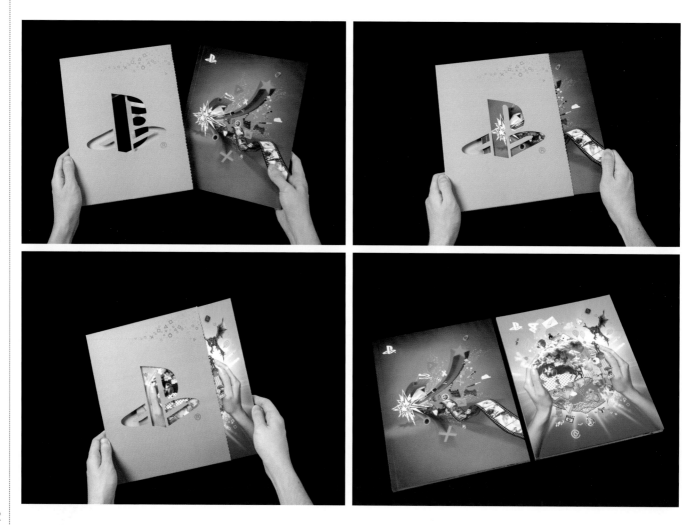

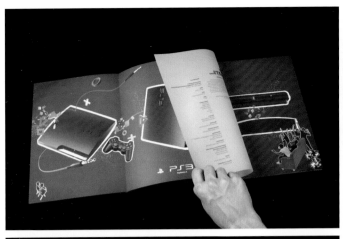

Treverket is a carpenter shop who is committed to quality and good materials, great joy and pride in their trade. They offer special decor and furniture, where they work closely with the customer.

Throughout the identity Ghost wanted to emphasize Treverket's quality and attention to detail. The logo has multiple levels and depths with small details where the inspiration came from moldings and materials. The designers have designed the logo, business cards, envelopes, letterheads and a new website.

Identity for Treverket

Design Agency: Ghost

Client: Treverket

Printing technology: digital 4-color process (CMYK) laser printed on wood

Printing technology: foil stamping

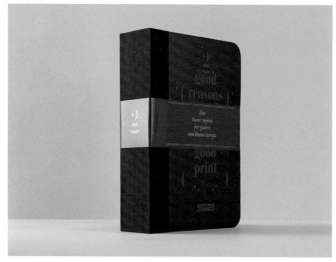

If you need two good reasons to enjoy a good print, check out this event packaging for a renowned Italian printing company.

SIZ Bottle Box

Design Agency: Happycentro
Client: SIZ printing company

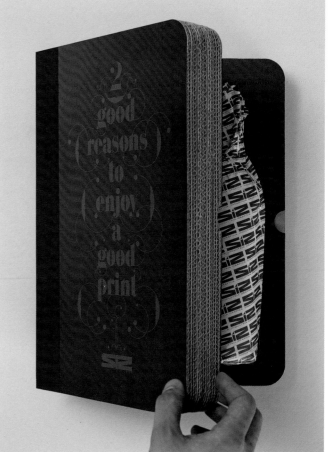

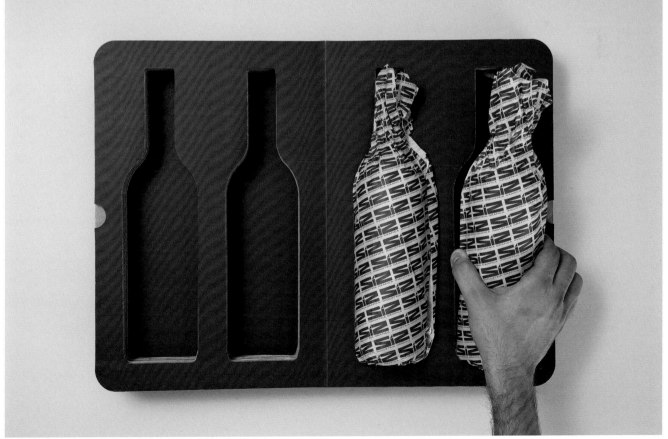

A self initiated project to produce a typographic deck of playing cards. Produced in partnership with paper and print company.

All 52 cards and jokers are recreated using unadulterated typefaces, avoiding any repeated fonts. The cards become a typographic primer. The project is produced as a set of cards and a limited edition poster.

Deck

Design Agency: hat-trick design

Client: Jim Sutherland

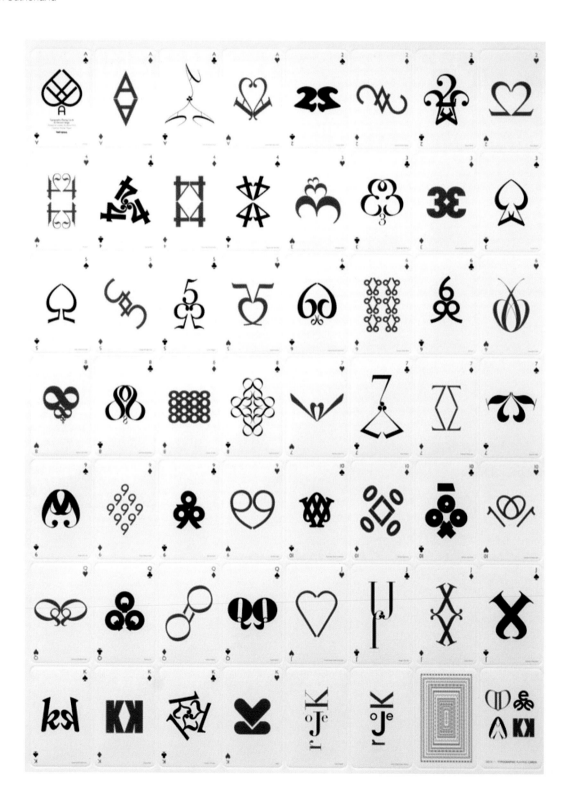

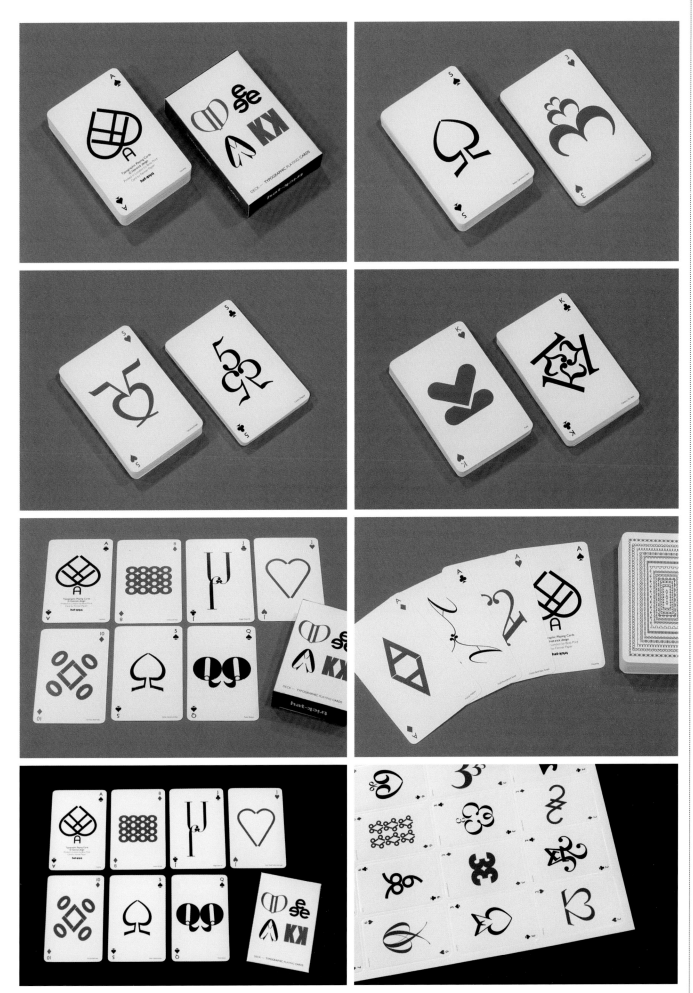

Printing technology: litho printed and blind embossed on McNaughtons Skye silk 350gsm

For 250 years, the Royal Botanic Gardens, Kew has led the world in plant and fungal science. And today, with hundreds of scientists at the centre of a global network, they are leading the quest to find plant-based solutions for people and the planet.

The design is a map of the world made from leaves — flattened by dividing its surface into gores (tapering segments). The curves are calculated so that when the points meet at the north and south, the map turns into a sphere. This map has been overlaid with a leaf skeleton to form a breathing planet. The segments became a strong graphic device that has been used across other communications with the campaign.

Through the breathing planet campaign, Kew are seeking to raise £100 million to strengthen themselves and to help build a more resilient planet.

Kew Breathing Planet

Design Agency: hat-trick design

Designer: Jim Sutherland, Gareth Howat + Alexandra Jurva

Client: Kew Foundation

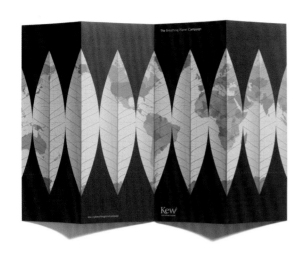

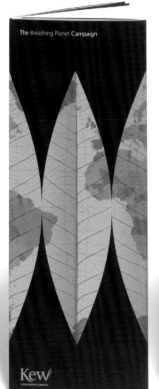

Printing technology: litho print on Antalis McNaughton Challenger offset 300gsm (Text + Cover) 140gsm (Dust jacket)

The catalogue uses sculpted embossing.

Catalogue

Design Agency: Joseph Rossi / Graphic First Aid

Client: Teracrea

Business cards were printed on a triplex smooth Colorplan board manufactured by GF Smith to reach 1050 gsm. Each graphic element was printed by IST printing services using a debossing and foil blocking stamping process on both sides of the business card.

DMWORKOOM Business Card

Design Agency: DMWORKROOM

Designer: Denis Mallet

Client: DMWORKROOM

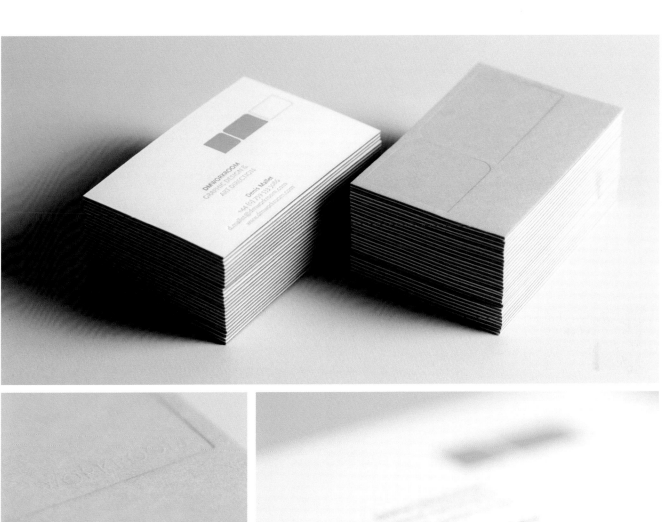

Printing technology: special paper, gold foil blocking, fold, lamination

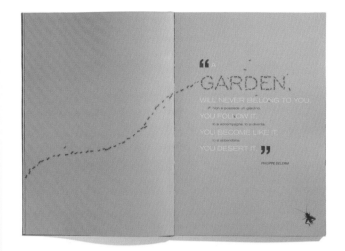 

 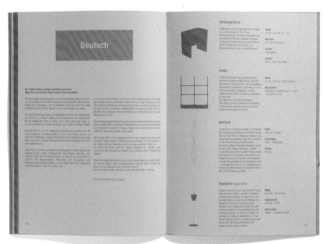

The dark brown folder looks solemn.

Folder

Design Agency: Joseph Rossi / Graphic First Aid

Client: Cariolato

Printing technology: special paper, gold foil blocking, fold, lamination

The press palette has used the dark tone.

First Press Palette

Design Agency: Joseph Rossi / Graphic First Aid

Client: Alcantara

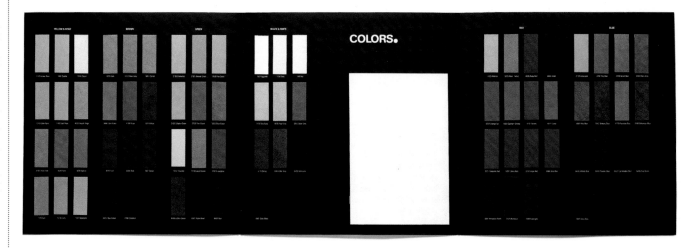

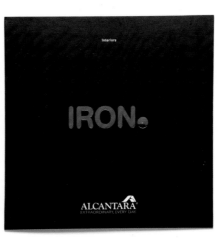

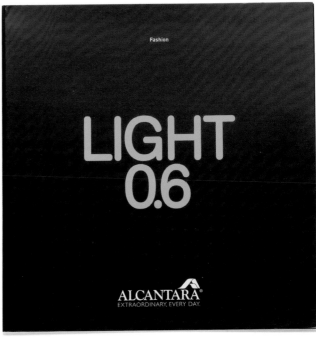
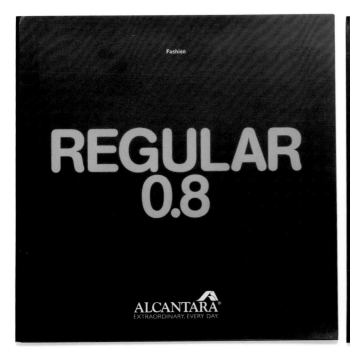
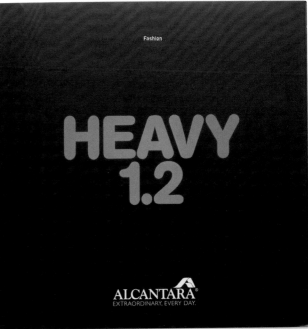

It is clean and neat.

Invitation

Design Agency: Joseph Rossi / Graphic First Aid

Client: Alcantara

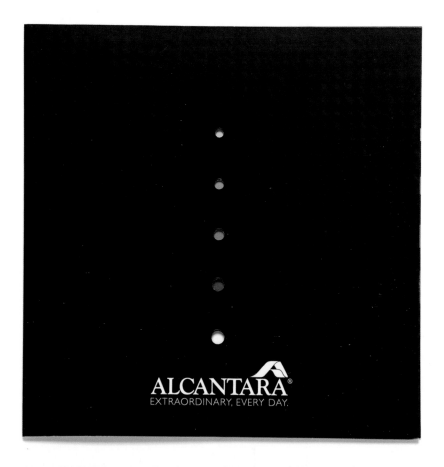

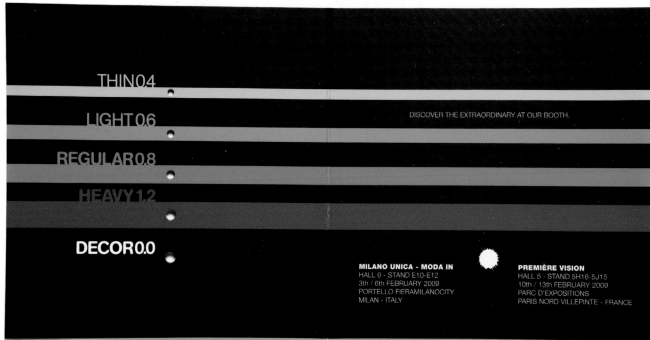

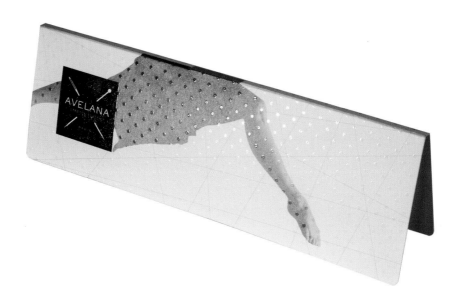

The project has used impressive patterns and printing techniques.

Header

Design Agency: Joseph Rossi / Graphic First Aid

Client: Avelana

 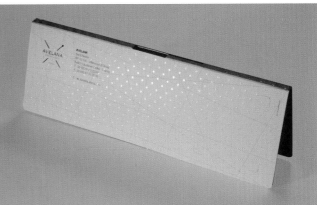

Printing technology: solid color gold, UV, fold

Printing technology: pantone, UV spot, "Pilke" paper touch, serigraphy on Alcantara, matt lamination on the box

There are more changes in the press palette than others.

Second Press Palette

Design Agency: Joseph Rossi / Graphic First Aid

Client: Alcantara

Printing technology: pantone, UV spot, "Plike" paper touch, serigraphy on Alcantara, matte lamination on the box

Dream   Lovely   Cloud

A Wadded Dream's handcraft makes use of cotton to create the sky, the sky in blue and white, the sky with sheep, elephant, horse and her dreams. Make use of cotton, she creates animal like accessories.

To create and consolidate their brand identity, they made use of set of stationary, postcard and lovely cushion package on their animal designed accessories.

The designers wish to share handcraft's sky and dream with all of their customer, and day dreamer.

A Wadded Dream

Design Agency: THINGSIDID

Designer: Kevin Ng & Vanessa Chan

Client: A Wadded Dream

棉襖之夢

A   Dream

Lovely   Cloud

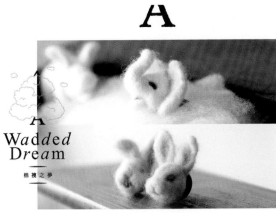

Printing technology: screen printing & stamp printing

053

Printing technology: screen printing & stamp printing

Printing technology: screen printing & stamp printing

Company profile of Tipografia San Martino developed on the topic of environmental sustainability. Natural elements' issue, designed with a pop-up use, draws the company profile in a geometrical way. Each page has a different technique, so can introduce services that typography offers.

Company Profile of Tipografia San Martino

Design Agency: Kalimera

Designer: Luca di Mira, Alessandro Marani

Client: Tipografia San Martino

Printing technology: pop-up, relief printing, ridgy UV-ink, paperboards FSC, laser engraving, embossing and debossing, foil stamping, coating and glazing, laminating, fluorescent ink, cutting and folding

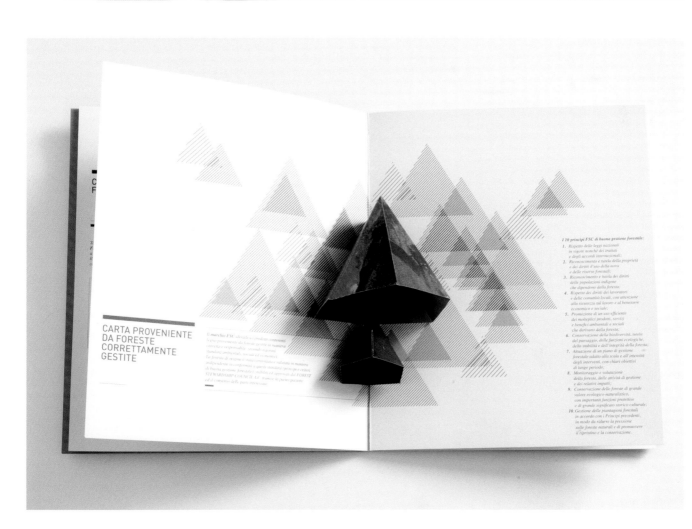

Printing technology: pop-up, relief printing, ridgy UV-ink, paperboards FSC, laser engraving, embossing and debossing, foil stamping, coating and glazing, laminating, fluorescent ink, cutting and folding

Printing technology: pop-up, relief printing, ridgy UV-ink, paperboards FSC, laser engraving, embossing and debossing, foil stamping, coating and glazing, laminating, fluorescent ink, cutting and folding

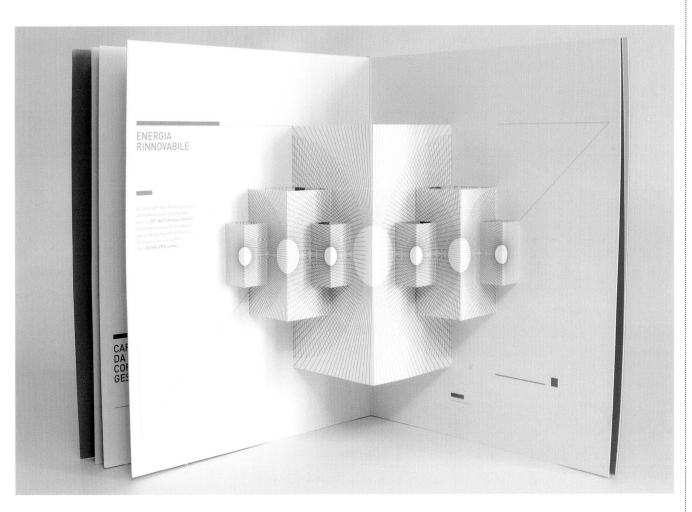
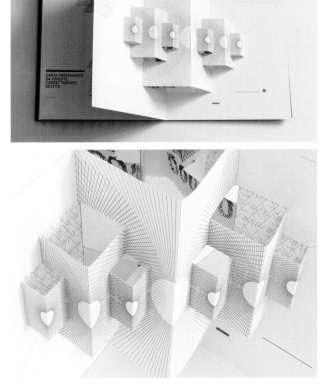

Printing technology: pop-up, relief printing, ridgy UV-ink, paperboards FSC, laser engraving, embossing and debossing, foil stamping, coating and glazing, laminating, fluorescent ink, cutting and folding

Pointpath Studios wanted a set of business cards that would stop recipients in their tracks. To that end, Seamless Creative developed the original design as an extension of the Pointpath brand, using the contrast of black and neon colors to give the cards punch. They then worked closely with their printer Mama's Sauce to develop a production method that would push the design over the top, by printing black, white and clear inks on top of triplexed neon paper. The result is exactly what the client wanted — a striking, memorable card that will spark curiosity and conversation.

Pointpath Studios Business Cards

Design Agency: Seamless Creative

Designer: Courtney Eliseo

Client: Self Promotion

Contributor: Mama's Sauce

Black Umbrella builds practical, efficient disaster plans for individuals, couples and families. The client chose the name umbrella for its symbolism of protection. The designers wanted to build upon the idea. As most of their services are for an individual's or families' home, the designers used the outer shape of an umbrella as a shield for the house. The logo can be used on its own or can be combined with a complex grid which represents the intricate network of connections and services.

Black Umbrella Stationery

Design Agency: MyORB

Designer: Lucie Kim, Felix von der Weppen

Client: Black Umbrella

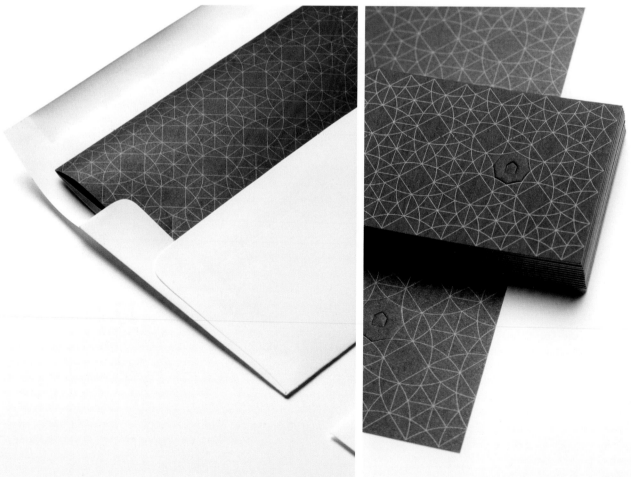

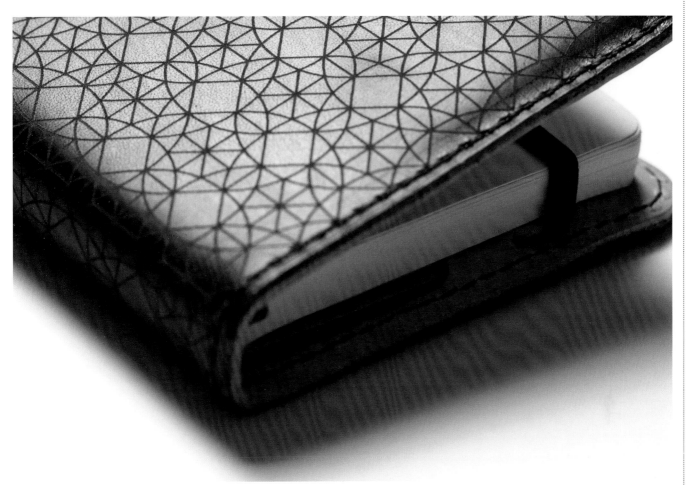

Printing technology: offset, blind emboss, silver ink

065

Printing technology: offset, secondarily monoprints, wax embossing, stamping

The patterns were made with mono-prints of tree trunks and leaves and combined with a geometric line pattern taken from the logo lines.

The whole package consists of a hand printed CD, a paper case sewn with black thread and wrapped with a two color offset, double-sided poster and sealed with white wax. The poster after it's opened can be used as a real poster signed and numbered by the band.

Mockbirth EP

Design Agency: Nevertheless

Client: Mockbirth

Printing technology: offset, secondarily monoprints, wax embossing, stamping

Printing technology: offset, secondarily monoprints, wax embossing, stamping

Fa!rbar is a non-profit bar driven by volunteers. The designers' approach to the assignment was to create a corporate identity which took its conceptual approach inspired by 60's wallpapers cosy living room. The wallpaper are made of rosettes with different motives referring to activities in the bar.

Fa!rbar

Design Agency: OddFischlein

Designer: Gudjon Freyr Oddsson/ Klaus Matthiesen

Client: Fa!rbar

The literary work "Eyes on Hands on" collects the Aarhus-based artist Jette Gejl Kristensen's experiences and works through the past 10 years.

The concept is based on Jette Gejl Kristensen's method and systematic approach to exploring the virtual and very sensuous medium. The desire is to present the viewer with a system which at first glance does not reveal the actual works, but requires interaction from the reader (flipping, turning, warming up varnish, breaking perforations etc) before each "virtual room" can be "opened". Each chapter has its own format and contains various works and studies.

Eyes on Hands on

Design Agency: OddFischlein

Designer: Gudjon Freyr Oddsson/ Kim Lange/ Klaus Matthiesen

Client: Eyes on Hands on

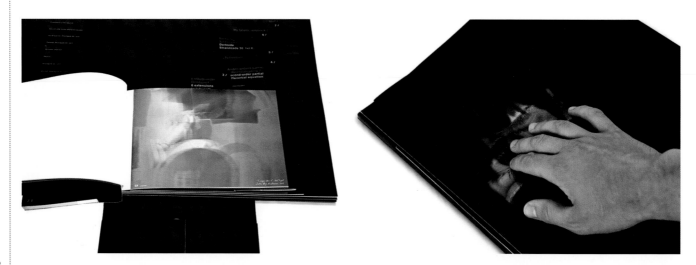

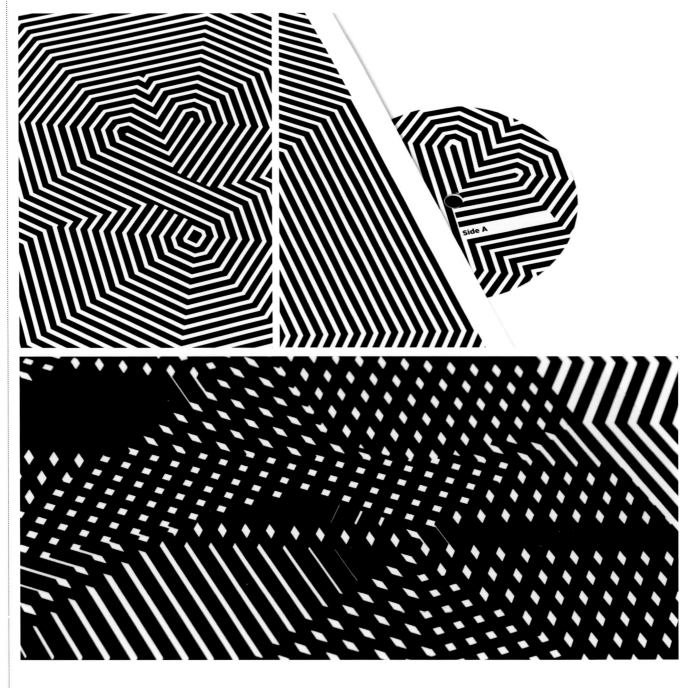

The visual concept reflects the meeting between the mechanical and the organic elements.

The intention is to influence the viewer visually, so that you not only play with the cover, but also get a sense of uncertainty when looking at it.

The concept incorporates the experience of many layers which you can hear in MachineDear's experimental and designed musical universe.

The vinyl is wrapped in a discobag which is pushed into transparent PVC. A pattern is printed on the discobag and PVC. When the discobag is taken out of the PVC, the gap between the patterns creates a constant change in the pattern's appearance.

MachineDear

Design Agency: OddFischlein

Designer: Klaus Matthiesen / Gudjon Freyr Oddsson

Client: Machine Dear, Killing Something that is allready dead

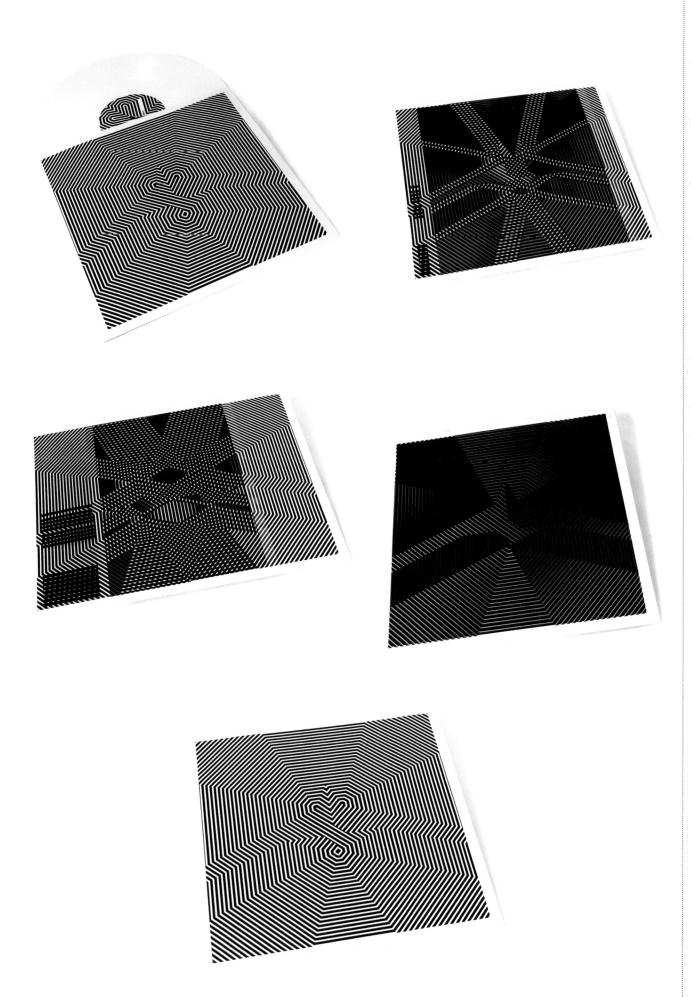

Printing technology: discobag is offset print; the PVC sleeve (screen print)

It is the VI design for the pub.

Kitashinchi Bar

Design Agency: otto design lab.

Designer: Takahiro ichino

Client: Kitashinchi Bar

Foil printing and custom-made envelopes were used to create an elegant, celestial-inspired invitation for EO's "Night of the Stars" — a Gala Awards dinner that celebrates for the top local entrepreneurs.

EO Invites

Design Agency: Nicework

Designer: Rowan Toselli

Client: EO (The Entrepreneurs' Organization)

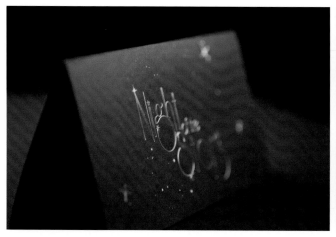

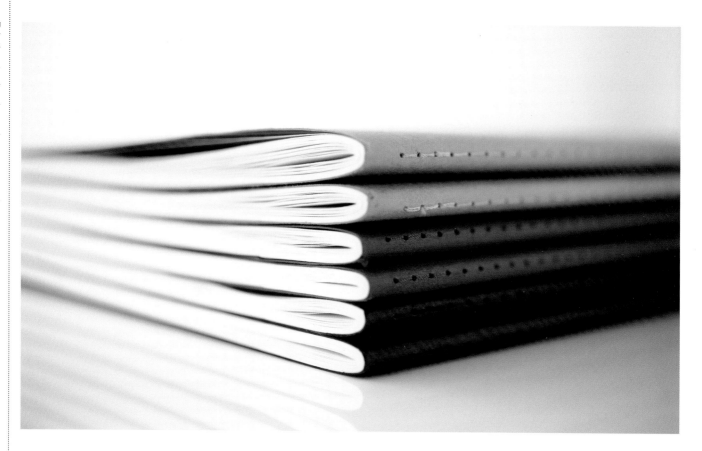

These notebooks are personal projects to test the laser engraving technique and have some fun. The process began with hand-drawn illustrations which were then translated into vector designs, with the end product being these engraved, recycled paper-bound books.

Paper Bound Engravings

Designer: Rowan Toselli

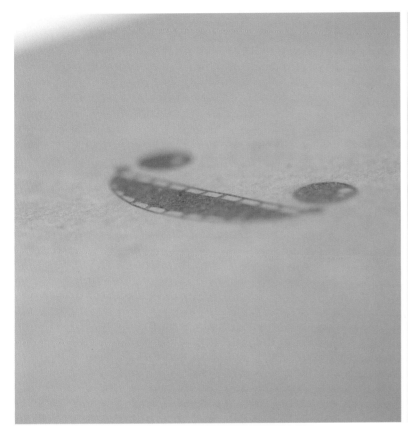

The wooden box was made from plywood and laminated with an 80-year-old Indian rose wood veneer. There is a black cardboard tray that slides in and out of the box which contains a tray with three leave-behinds: a digital copy of the designer's portfolio, a small wooden badge/button, and a laser-engraved illustration onto a piece of jacaranda wood. And finally the designer's printed portfolio rests on top of the inner tray.

Wooden Portfolio Box

Designer: Rowan Toselli

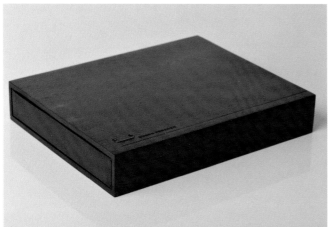

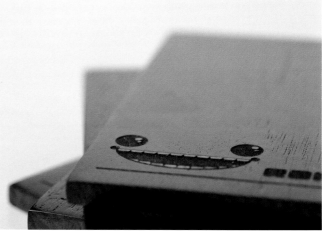

Printing technology: laser engraving

083

Sergio Mendoza Studio had the chance and responsibility of taking care of the complete project: from the interiors to the food, brand... Even had to go flea market hunting for croquery and vintage furniture!

They said it was fantastic to have clients coming in to say hello even before the works had started. There was the chance to meet the neighbors, the people and their dogs... This was very helpful since budget was extremely limited and there was only one month from the original idea to the actual opening!

So DIY was the deal. They were all the woodwork and kept it simple, cozy and friendly. The same idea was applied to the banding and which was printed. They designed a system and a "set of tools" to be used, so everything was actually co-produced by the staff.

La Canya

Design Agency: Sergio Mendoza Studio

Designer: Sergio Mendoza

Client: La Canya

The dark tone makes the calendar look cool.

Calendar

Design Agency: Joseph Rossi / Graphic First Aid
Client: Studio Vesco Giaretta

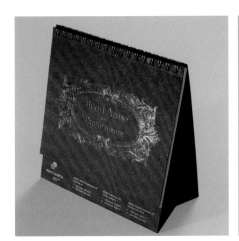

For this particular print, The Palmetto Press was challenged by accepting free range from both Adrian and Kim. From the design to the belly-bands, to the envelopes, this team put forth their best effort to create something beautiful for a dear friend. They are printed with a Chandler & Price letterpress using yellow, grey, and blue inks. This piece also included a wood grain blind impression for added texture and detail. The total came to 100 invitations, direction cards, response cards, belly-bands, and envelopes. Each direction card was die-cut, whole punched, and attached to a loop of twine. The envelopes were designed and cut out on a Kongsberg cutting table. The invitations were put together one by one.

Kim & Adrian Wedding Suite

Design Agency: The Palmetto Press

Designer: Jessie Thompson

Client: Kim & Adrian

Every year the BASF Coatings GmbH invites clients from the automotive industries to a presentation of their color collection. For this color show a special theme has to be conceived that always accords to the annual Global Trend Book Theme, which in 2011 was "Come Closer". For the Show 2011 the attendees were invited as "Come Closer Club" members — a fictional private club for honorary. Based on this concept invitations and event accessories were designed. Every member also received a club box with vessels and was requested to collect favorite items with great color impressions to discuss at the workshop.

Come Closer Club

Design Agency: ARE WE DESIGNER

Designer: Daniela Kempkes, Janina Braun, Ruth Biniwersi, Sascha van den Bloock

Client: BASF Coatings GmbH

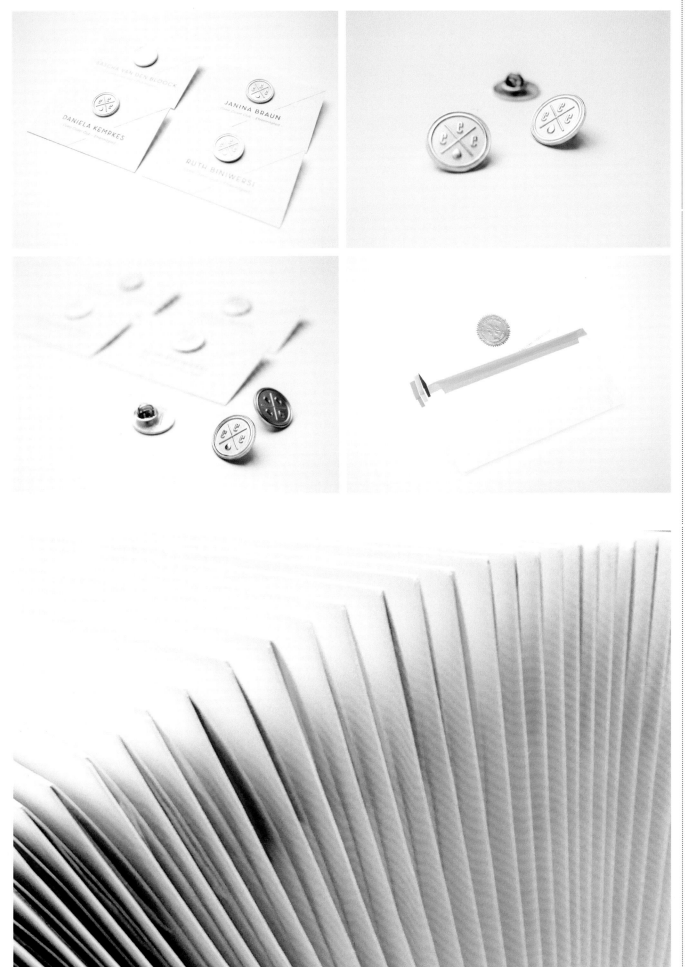

Printing technology: embossing on paper and golden stickers, stamping, envelopes with handmade tissue paper inlay

Printing technology: embossing on paper and golden stickers, stamping, envelopes with handmade tissue paper inlay

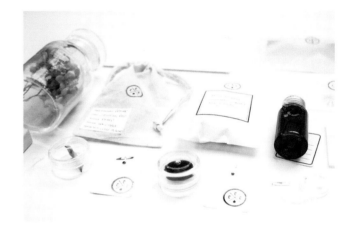

Printing technology: embossing on paper and golden stickers, stamping, envelopes with handmade tissue paper inlay

Printing technology: rubber stamping and offset, burned by using a branding iron

---

The branding created for the Finnish designer Harri Koskinen and his own furniture collection, works, is practical and innovative like the designer himself.

Harri Koskinen

Design Agency: BOND

Designer: Jesper Bange

Client: Harri Koskinen

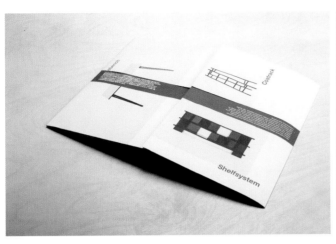

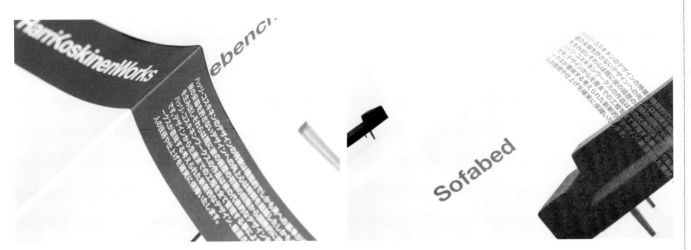

Printing technology: rubber stamping and offset, burned by using a branding iron

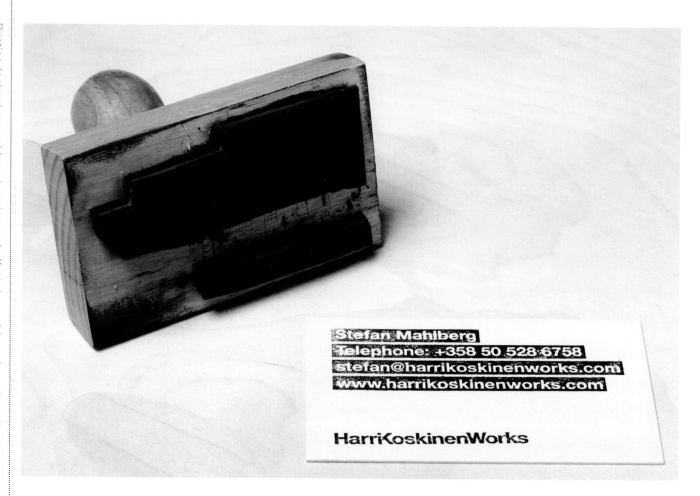
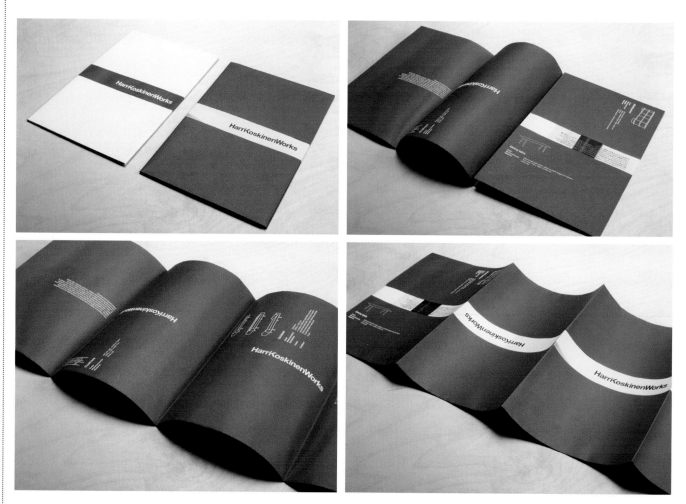

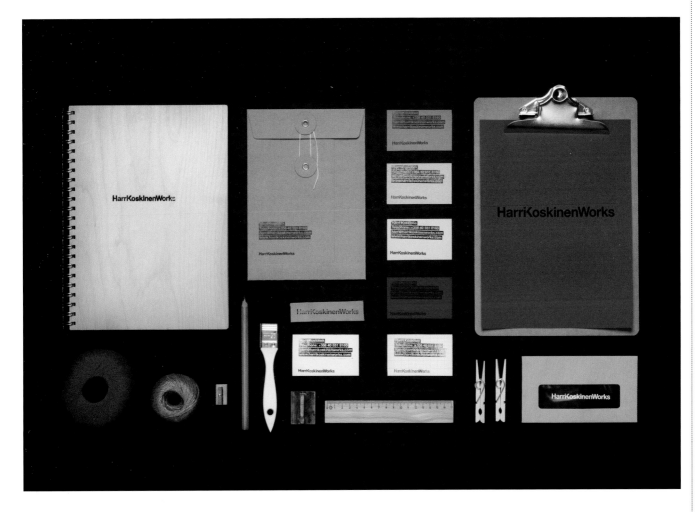

Printing technology: rubber stamping and offset, burned by using a branding iron

097

Meier Seefeld is a women´s clothing boutique set in the middle of the Tyrolean mountains.

The identity was designed to convey understated luxury in a comfortable atmosphere to both visitors and locals. The notion of traditional alpine winter sport chic is combined with a modern perspective on fashion and the zeitgeist of the village Seefeld.

Meier Seefeld

Design Agency: Bureau Rabensteiner

Designer: Mike Rabensteiner, Isabella Meischberger

Client: Ernst Meier

Printing technology: hot foil stamping, with black stickers embossed

Mangoola is a coal mine situated in New South Wales, Australia.

To celebrate the opening of the new mine, Xstrata Coal sought out End of Work to create launch and identity materials. The concept was based on the strata layers of the earth and that Australian Native bird the Black Cockatoo was found in the area.
Using the layers of the earth and the Black Cockatoo as inspiration studio End of Work created beautiful crafted print collateral to represent the feathers of a bird and the strata layers of the earth. This was achieved through a large foil blocked pattern stamped onto black paper, embossing and laser cutting paper.

Mangoola

Design Agency: End of Work

Designer: Designer: Justin Smith, Goran Momircevski and Ralph Kenke

Client: Xstrata Coal

Printing technology: foil stamping, offset, laser cut for the dots, popup

Printing Technology: silk-screen, embossed, cutting and folding

Reduction of materials such as ink, water and paper reduce carbon emission. Besides, it saves money wasted on mailing. Reducing is the fastest way to reach simplicity. Reduce your printings but with the best quality ever. Founded on these thoughts, Hélio Rosas company graphic identity, the H®, was developed.

Graphic Identity

Design Agency: HLD / hyperlocaldesign

Client: Helio Rosas

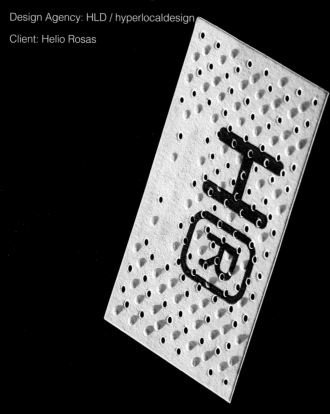
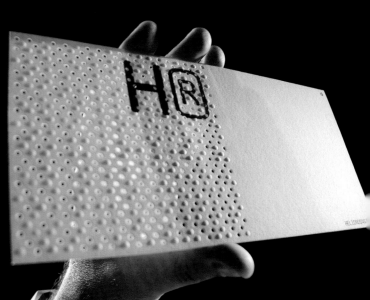

Printing technology: silk-screen, embossed, cutting and folding

103

Printing technology: silk-screen, embossed, cutting and folding

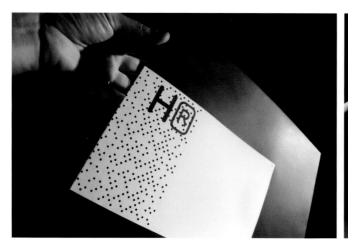
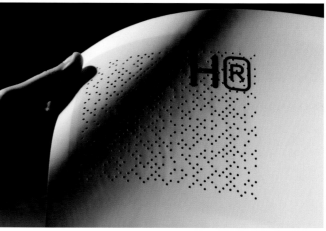
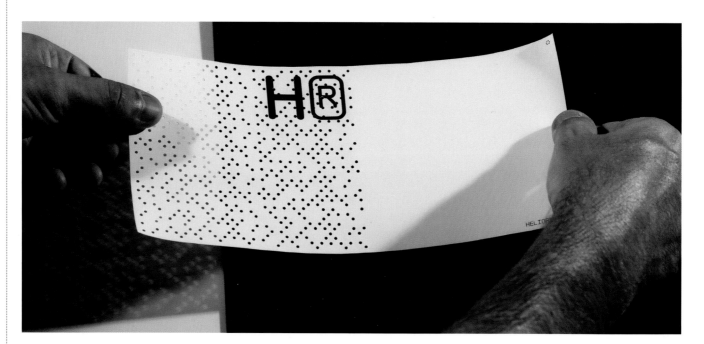
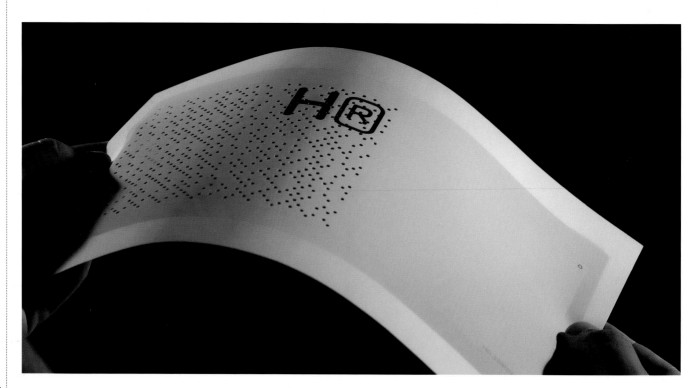

104

Arkigram is an architect agency, highly specialized in course management and technical solutions for the Danish architectural industry.

Arkigram — Brand Identity

Design Agency: Ineo Designlab®

Designer: Sebastian Gram

Client: arkigram

Printing technology: UV – inks (UV-inks are used on envelopes), offset printing

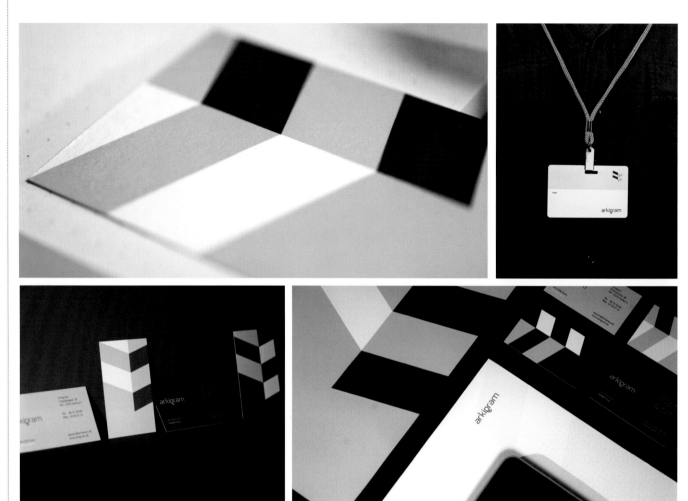

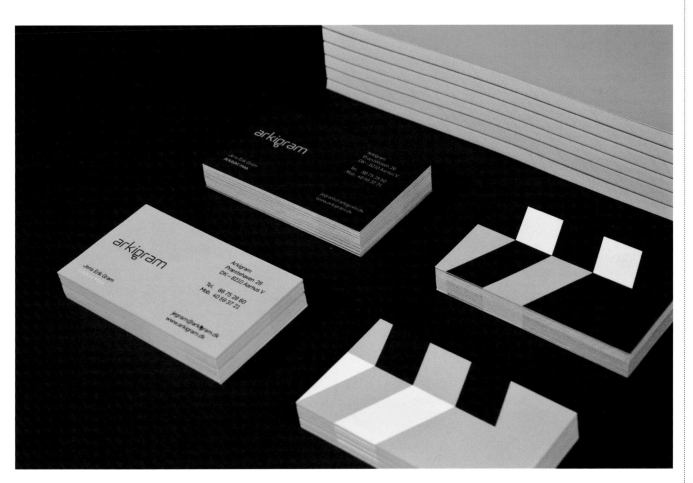

Printing technology: UV – inks (UV-inks are used on envelopes), offset printing

Las Caglias is a small restaurant in the Swiss Alps. Under the motto "the best things in life are the simplest" this small exclusive restaurant serves simple, locally interpreted dishes from classic Italian cuisine to quality-conscious local guests.

The name originates from the retro-Romanesque (the original, now rarely used, Swiss Alps dialect) where Caglias can be translated as "bush" — a reference to the restaurants beautiful location.

Las Caglias

Design Agency: Ineo Designlab®

Designer: Peter Christensen, Soren Herold

Client: Las Caglias

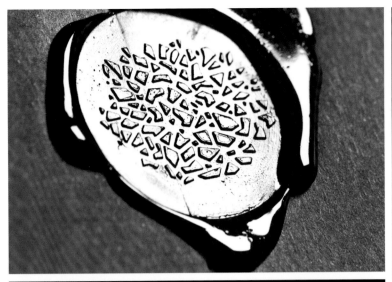

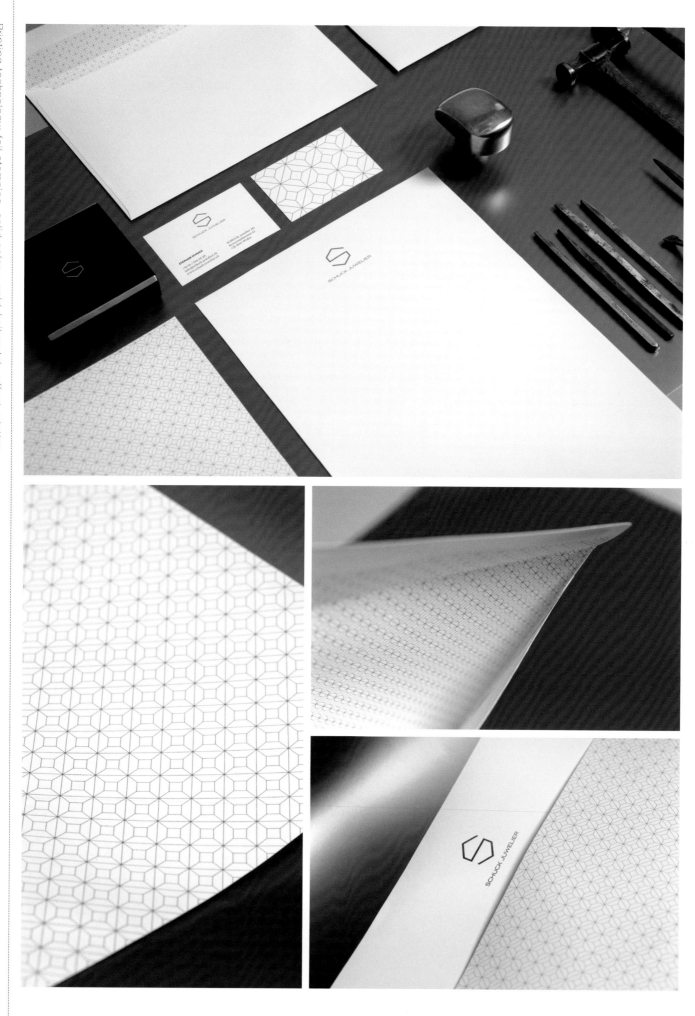

Printing technology: foil stamping, solid color – gold / silver ink, offset printing

Schuck Juwelier is a high-end jeweler based in Widen, Switzerland. Schuck Juwelier specializes in designing and producing unique, exclusive jewelry to a wide range of international clients.

Schuck Juwelier — Brand Identity

Design Agency: Ineo Designlab®

Designer: Peter Christensen, Soren Herold

Client: Schuck Juwelier

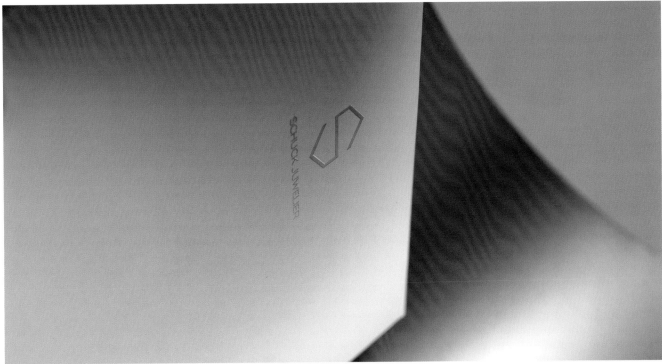

In 1958, Jacob Jensen started his design studio at the kitchen table of his home in Denmark. Over half a century later, Jacob Jensen Design opened the doors to Asia.

On the 3rd of August 2012, the first DeTao Master studio, Jacob Jensen Design I DeTao Shanghai was official opened in the CCIC building (Cultural and Creative Industry Cluster) on the campus of the Shanghai Institute of Visual Arts, part of the prestigious Fudan University.

The DeTao Masters Academy is an international training environment under the program of DeTao Master Heritures (DMH) launch by the esteemed DeTao Group.

The Opening of Jacob Jensen Design I DeTao Shanghai

Design Agency: Jacob Jensen Design

Designer: Nigel Hopwood

Client: Jacob Jensen Design I DeTao Shanghai

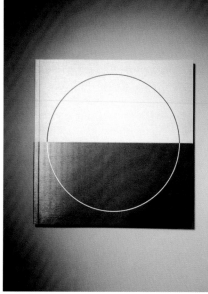

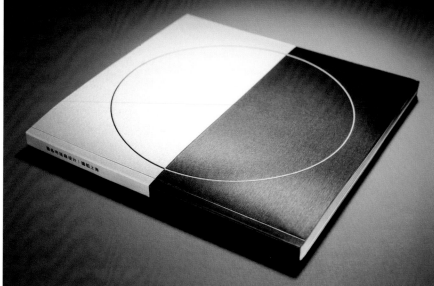

Printing technology: screen print, blind emboss, protection varnish (matt)

This card set depicts 6 well known Grimm Fairy Tale characters. All 6 are shown trapped in the predicament of their story to create a connection between their narratives. The folkloric feel and illustrative style of the images are underlined by the texture of the letterpress and the bold effect of the gold foil.

Fairy Tale Collection

Design Agency: Karolin Schnoor

Printer: Bison Bookbinding

Groovewear special edition box contains a limited edition tee shirt and silkscreen by famous European street artists. For each issue fine papers, special printing techniques and custom numbers.

Groovewear Artist Box

Design Agency: La Tigre

Client: Groovewear

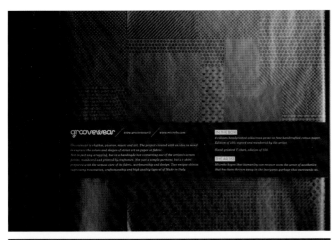
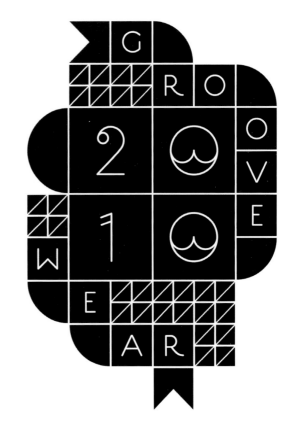

Many couples have a song that links them. In the case of Patricia and Gian this link is Jack Jonshon's song "Better togheter". The idea behind the invitation is to communicate the wedding details playing with the chorus of this song. A funny and unusual method was thought for discovering the details of the wedding. The guests themselves reveal the final message via scratching-off the golden ink. The wedding details thus appear interspersed in the chorus of the Jack Johnson's song.

Wedding Patricia & Gian

Design Agency: Raquel Quevedo in collaboration with Diego Ramos

Designer: Raquel Quevedo

Client: Patricia Solarte

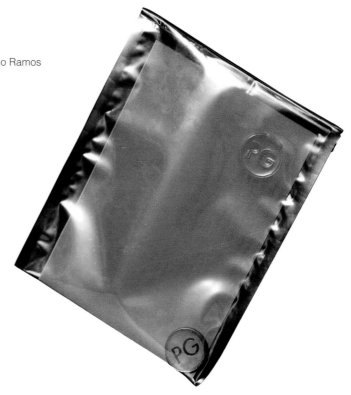

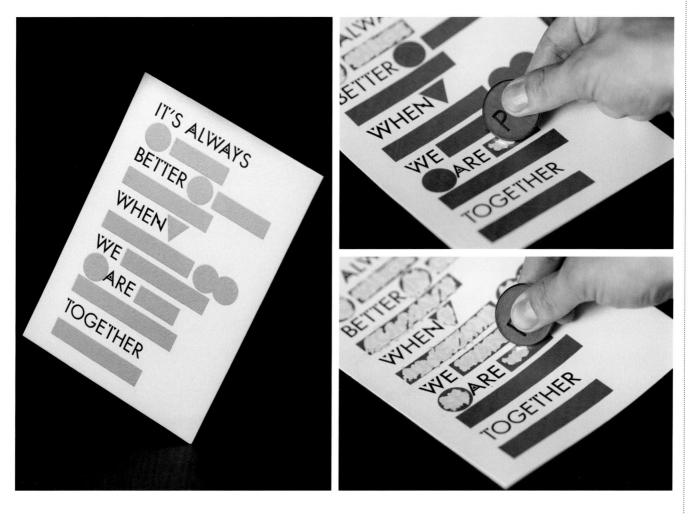

Printing technology: screen printing (green pantone), scratching-off golden ink, UV-ink, embossing in the envelope

To create a visual identity for the new Chinese high-end mens fashion brand, "M-IDEA FOREVER", based on a strong concept, which expresses the brand's two fundamental characteristics: Avant-Garde and technology.

The Carbon concept contains and supports the brand values and can on several different levels of abstraction, be used as the DNA and source of inspiration for the creation and evolution of the whole identity.

Apart from the fact that the element Carbon forms the basis of all known life on Earth, it is exceptionally because of its diversity.

Under different influences from its surroundings Carbon can take various physical forms, each of which possess completely unique and diverse properties, although they are identical in their chemical composition. Looking at the raw graphite and polished diamond, it is fascinating to imagine that they are both composed of the element Carbon.

Just like Carbon, the brand M-IDEA FOREVER is characterised by its diversity. The new brand is based on a unique style composition, which mixes poetic Avant-Garde and functional Technology.

Although it is very different and sometimes even opposite, the two elements of M-IDEA FOREVER tied together by the desire to constantly experiment play together and strengthen each other.

The black and soft Avant-Garde is expressed through the raw graphite. The bright and hard technology grows out of the polished diamond. The Carbon Concept unites them.

M-IDEA FOREVER

Design Agency: Scandinavian DesignLab

Designer: Per Madsen, Line Arlander, Christian Steen Jörgensen

Client: Mark Fairwhale

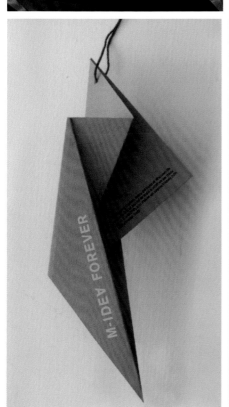

Printing technology: cutting and folding, foil stamping, embossing and debossing, solid color – gold / silver ink, offset printing

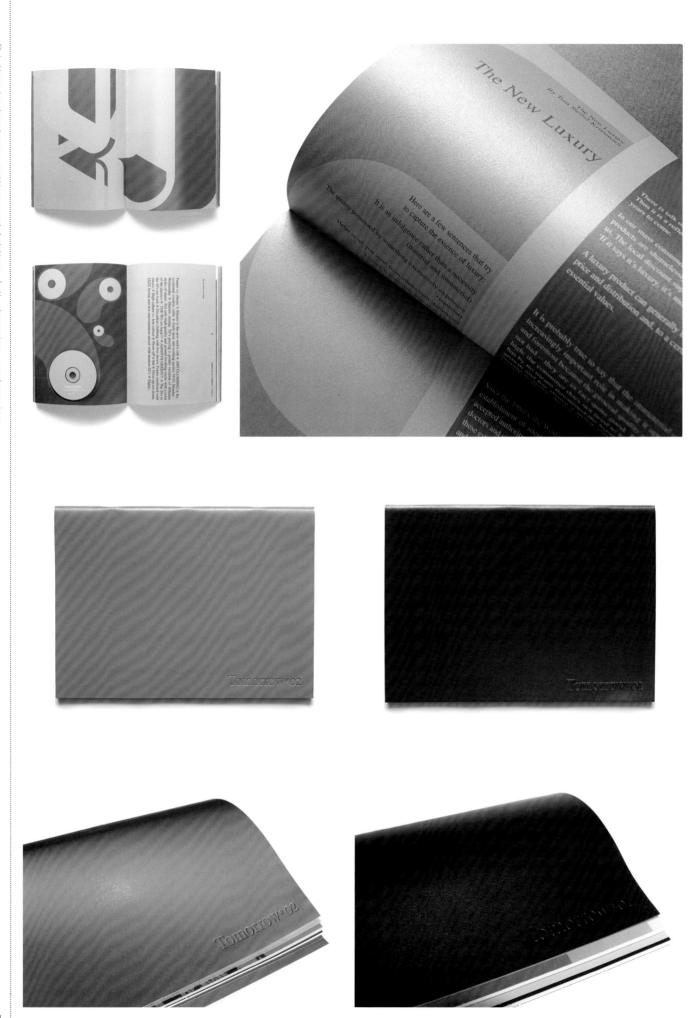

The task was to create a twice-a-year published trend book for internal use among designers and creative artists. The book which both in shape and content can be used as a source of inspiration on trends in music, fashion, retail, art and sociology.

Just like a chameleon the publication maintains its shape, but constantly change appearance to fit into the surroundings. As the edition is limited a special hand-made bookbinding is chosen, which gives unlimited possibilities to experimentation with new paper qualities and other materials like leather. This makes it possible to develop a unique product that offers the user a new sensual experience according to the overall theme through the many different articles.

Tomorrow

Design Agency: Scandinavian DesignLab

Designer: Per Madsen

Client: Bestseller

Printing technology: cutting and folding, foil stamping, embossing and debossing, solid color – gold / silver ink, offset printing

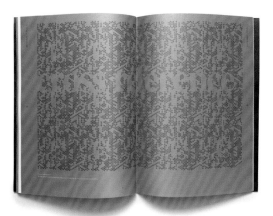

126

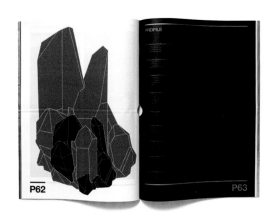

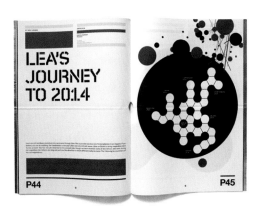

Printing technology: cutting and folding, foil stamping, embossing and debossing, solid color – gold / silver ink, offset printing

Printing technology: offset printing, silver ink

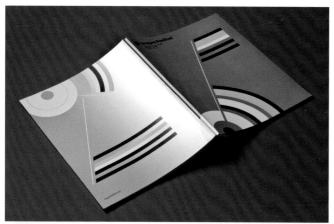

In 2012, it marked the start of 25th anniversary of the Images Festival, the largest festival in North America for experimental motion-based art. The studio was engaged to design all of the marketing collateral, including program guide, transit advertising, signage, T-shirts, passes and trailer. To celebrate the festival's many years of success while informed by the traditional film countdown, the visual identity featured a distinct set of numerals which was the designers designed for use in both static print and kinetic motion applications.

Images Festival 25

Design Agency: The Office of Gilbert Li

Creative Director: Gilbert Li

Designer: Brian Banton

Client: Images Festival

The gifts have blended in Chinese elements and contemporary styles.

Sabores Da Vida Gifts

Design Agency: Tong Design Studio

Designer: Wai Hang Tong

Printing technology: cutting and folding, UV-inks, offset printing, stone paper

Printing technology: rubber stamping, debossing, foil stamping

La Forma Saporita is a bachelor thesis that the designer has done in 2 months. It is supposed to be an Italian brand based in Parma that manufactures pasta goods (e.g. spaghetti, tagliatelle, rigatoni, etc.). The brand name means "the tasty shape" because the pasta comes in various shapes and it is always delicious.

When it comes to branding, the designers all know that print design suggests a large scope of action. You will find two versions of business cards (one for corporate clients and the other for the regular ones), fancy envelopes, paper, flyers, posters, a pocket notebook and a variety of stamps and seals. Wax seals and paper embossers, a classical means of protection, are not only gaining popularity nowadays but also give the brand a sophisticated look.

There are even more branded goods and accessories to be seen in this project but the most essential part is the package — a simple but at the same time sharp design. The glass which the package is made of, as well as the protecting cloth and the coated labels printed in metallic pantone contributes to the distinctive look.

All that pasta talk should have made you a little hungry but unfortunately La Forma Saporita is still a concept and is real far from the production line.

La Forma Saporita

Designer: Yanko Djarov

Printing technology: rubber stamping, debossing, foil stamping

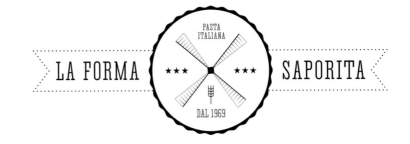

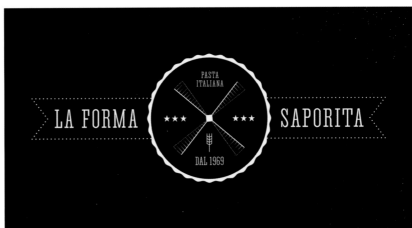

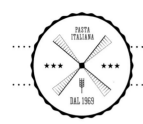 LA FORMA SAPORITA

FOR 3 PEOPLE

299G

1 2 3 4 5 6 7 8 9

WWW.LAFORMASAPORITA.IT

LA FORMA SAPORITA

WWW.LAFORMASAPORITA.IT

SITO UFFICIALE

★★★

HOME

ABOUT

PRODUCTS

RECIPES

DOWNLOADS

CONTACT US

★★★ PARMA, ITALIA ★★★

Printing technology: rubber stamping, debossing, foil stamping

Printing technology: rubber stamping, debossing, foil stamping

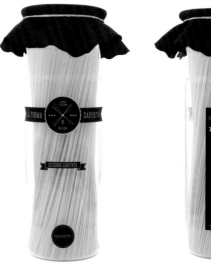

As a nice convergence of holiday mailer, self-promotion and poster, this handy poster comes with tear-off days so that you can be painfully reminded how long it is taking you to fulfill your resolution, which you can write at the top of the poster.

The poster is perforated 20 times across and 20 times down at very narrow intervals, giving the poster an oddly yet appealing fragility. Even more surprising is that the posters are hand-perforated with a "Carl" by the designers, as they explain, "no one would take the job fearing that they would make the poster look like a potato chip".

Perhaps an appropriate resolution for this poster would be "I will tear off all 365 days of this poster".

365 Poster

Designer: Arvi, David Muro II

This was a commissioned wedding invitation. The client told the desdigners she was having a fairy tale themed wedding, and gave the designers pretty much carte blanche to build something beautiful. This invitation went through a number of iterations and design concepts. The designers began with the idea of using Neuschwanstein Castle as a backdrop and evolved to a multilayer popup incorporating the silhouette of the couple, and a carriage. The original motions were quite a bit more complex, but the designers evolved it into a simple, modular assembly mechanism that they had adapted to use in other popup cards.

Fairy Tale Themed Wedding Invitation

Design Agency: Artifacture

Designer: Donna Hawk

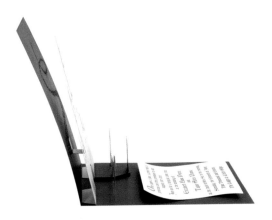

Printing technology: laser cutting, foil stamping, hand assembly, laser printed

This wedding was being held at a local boutique hotel with a large kinetic installation of meshed gears. The clients were Steampunk enthusiasts, and the venue had a very strong retro-industrial vibe. The designers explored static options, but the sculpture was such a great opportunity that they just could not pass it up. Simply putting the invitation on the moving gear was too much of a gimmick, so the designers hit upon putting openings in the body of the gear and dividing the event data into sections so that the gear motion was not only ornamental, but truly functional.

Kinetic Gear Invitation

Design Agency: Artifacture

Designer: Donna Hawk

Printing technology: laser cutting, printed with thermography

Printing technology: laser cutting, printed on an Indigo digital press

These invitations were done for the premier fundraising event of the local chapter of the Links Organization. The event celebrated the 20th anniversary of their black tie boxing event. The green boxing gloves are a trademark of the event, as was the round designation, so it was important that they be integrated. After presenting a number of different designs, they settled on this simple but elegant popup.

Pop-up Fundraiser Invitation

Design Agency: Artifacture

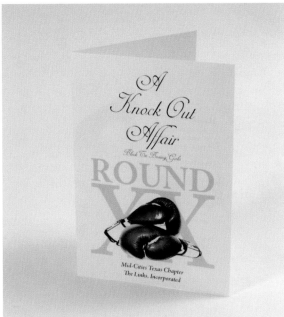

Graffiti is often misunderstood, and it is difficult for a graffiti writer to explain what he feels about it. Designer has been asked by a book publisher to write few pages to explain to teenagers how to practice graffiti. After hours and hours the designer came to the conclusion there was nothing to teach about it except providing the tool to do it. The new generation needs to discover it by practising it instead of reading lessons. Rather than a long sterile speech debating on the good or bad virtues, the designer created his own book to share his point of view through a conceptual message.

All You Need to Know about Graffiti Is in This Book.

Designer: Benoit Ollive

Printing technology: silver ink letter-pressed cover, hand-made binding, laser cut paper

Printing technology: silver ink letter-pressed cover, hand-made binding, laser cut paper

To be able to avoid what people learnt at school they first needed to follow a lot of rules, in order to understand how they work. Forgetting these rules allows you to be free, independent and most of the time it makes you feel unique. This poster is all about using the rules, the precision of the letter-press process, the choice of the paper and ink, the time spent on the work but it is also about destroying all these things to make a unique object.

Set of 10 posters. Letter-pressed handmade erasable ink on pure white glossy paper, signed and numbered.

Erasable Poster

Designer: Benoit Ollive

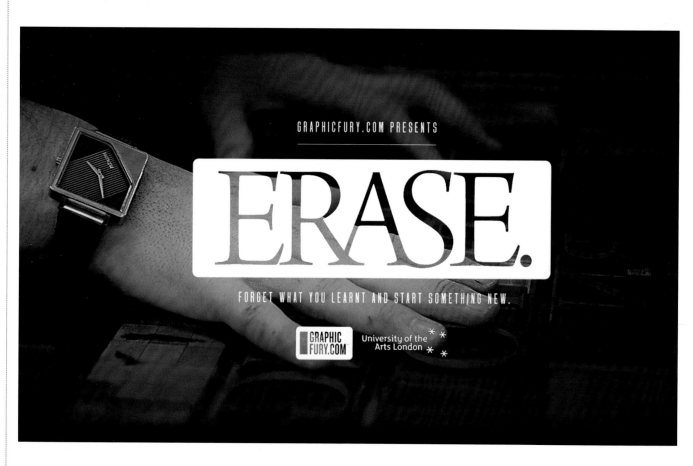

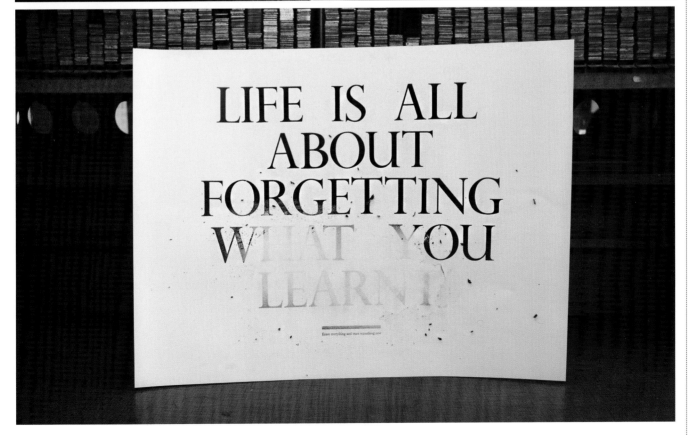

Printing technology: letter-pressed handmade erasable ink

Printing technology: silk screen onto golden metallic silk, to mimic the steel feeling of a skyscraper, with 4+2 colours + foil inside

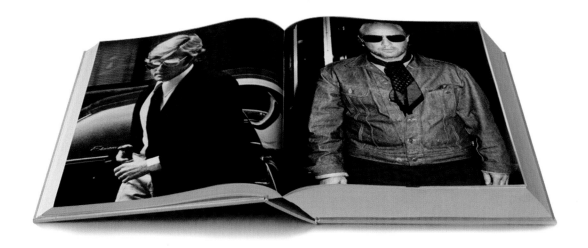

Become was briefed to direct the overall graphic approach to the book — a brief with a clear focus on designing an innovative luxury book case with the purpose of becoming a piece of luxury furnishing. Gloria wanted the bookcase to capture the city and it's grand architecture.

Henrik designed a vertical standing bookcase literally looking like a tall NY-building. A slipcase mimicking the glass of a skyscraper is covering the book, which is raised vertically sitting in a bronze coloured "steel" base.

New York-Limited Edition

Design Agency: Become

Client: Gloria Luxury

This collection of wedding stationery has been inspired by the art deco movement of the 1920's and 1930's. The liner symmetry echoes the designs of the period and is popular with couples who want to bring a touch of glamour and elegance to their wedding stationery. The suite is named Cleopatra to reflect the impact ancient Egypt had on the movement.

Cleopatra Wedding Stationery Suite

Designer: Eva Slade

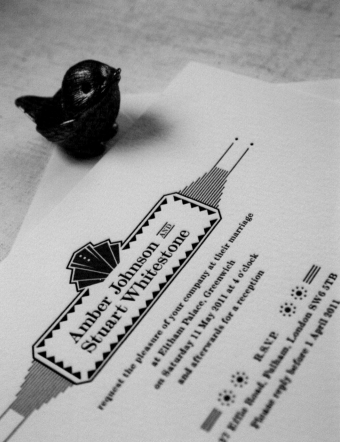

This wedding stationery suite called Eleanor has been designed for a wedding that is destined to be a relaxed and informal event. Playful and lighthearted is set the tone for a fun packed day.

Eleanor Wedding Stationery Suite

Designer: Eva Slade

This suite of wedding stationery called Marianne has been designed with retro flare in mind.

Marianne Wedding Stationery Suite

Designer: Eva Slade

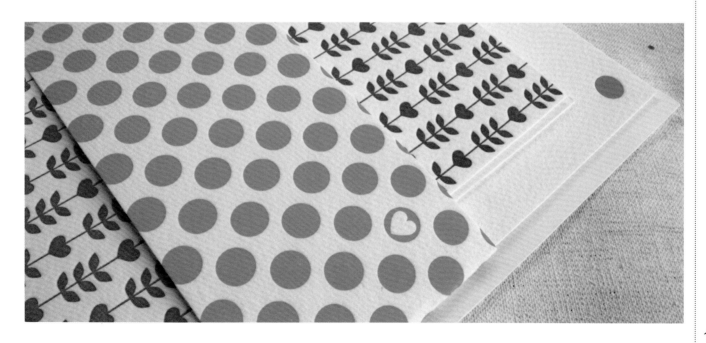

filthymedia strongly believe in the value of print, so as a print-based design studio. They felt it important that their stationary is a showcase of what they can do within the medium. filthy created a bespoke logo that was then embossed across all of the stationery, creating a latex texture on uncoated paper. filthy used a color palette set in black & white, with a hint of turquoise to demonstrate the attention to detail they like to achieve with all of their projects.

filthymedia Corporate ID

Design Agency: filthymedia

Designer: Joe Pilbeam

Client: Self Inititiated

Printing technology: triplexed color plan business cards, black foiled embossed logo, teal & white foil detailing, double hit black color plan envelopes, lamination

Printing technology: triplexed color plan business cards, black foiled embossed logo, teal & white foil detailing, double hit black color plan envelopes, lamination

Printing technology: letterpress printing (coasters), laser engraving (whiskey tumblers), rubber stamping (on packaging)

In 2012, it brought Firebelly many new opportunities and relationships with friends and clients. To show their appreciation, the studio crafted a personal holiday gift coaster set. As an extra "thank you" a few pairs of designed and engraved whiskey tumblers were included along with a select number of sets.

Using shapes, colors and motifs inspired by the season of cold, each set contained four die-cut and letterpressed, 2-color, hand-stamped edition coasters. Also included custom stamped muslin cloth wrap, tag and personal note. Each coaster had a friendly, glass-half-full message which included custom lettering and illustration.

Firebelly Holiday Gift

Design Agency: Firebelly Design

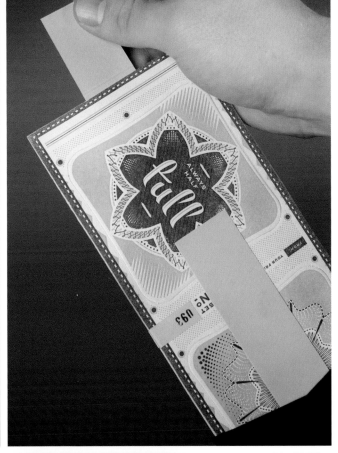

Printing technology: letterpress printing (coasters), laser engraving (whiskey tumblers), rubber stamping (on packaging)

Printing technology: offset printing, laser engraving, cutting and folding, embossing

The design is for the 10th anniversary of the famous Stuttgart Bar "Suite212". Theme was "Jail Hotel". The designers created a graphical identity by alienating the baden-württemberg national emblem. (As summonings are pink here, they had no choice with the color) Invitation was sealed with wax and showing the number "10" the key visual for their reverse-grafitti guerilla campaign watch out for the "10".

Invitation printed on pink paper, wrapped by natron kraft paper and sealed by wax (showing the "10"). Inside there had been printed the map of the location.

10 Jahre Suite212

Design Agency: LSDK

Designer: Christian Voegtlin

Client: Gastromedia Betriebs GmbH – Suite212

Copywriter: Sergej Grusdew

Printing technology: offset printing, laser engraving, cutting and folding, embossing

Corporte brochure for luxurious apartments that preserve the history of the early 20th century building in contrast with a modern interior design and the latest technology.

El Palauet Living Barcelona / Corporate Brochure

Design Agency: Marnich Associates

Designer: Wladimir Marnich & Anna Sodupe

Client: El Palauet Living Barcelona

Printing technology: foil stamping, screen printing

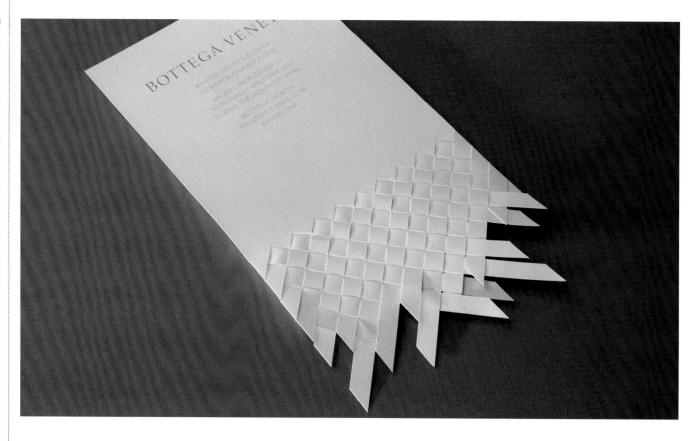

It is an invitation design proposal for the opening of Bottega Veneta's flagship store in Barcelona.

Bottega Veneta

Design Agency: Marnich Associates

Designer: Wladimir Marnich, Susana Catalan and Griselda Martí

Client: Equipo Singular

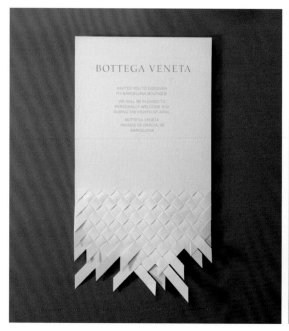

Catalogue for Matali Grasset's exhibition at Rabih Hage Gallery.

Matali Grasset

Design Agency: Marnich Associates
Designer: Wladimir Marnich & Iris Tàrraga
Client: Rabih Hage Gallery, London

Corporate Identity for Noguera & Vintró, the gifts and stationery distributor.

Noguera & Vintró

Design Agency: Marnich Associates

Designer: Wladimir Marnich & Iris Tàrraga

Client: Noguera & Vintró

Printing technology: gold foil stamping

Corporate identity for a hotel based on "a little bit of rough in a luxurious London".

Rough Luxe Hotel

Design Agency: Marnich Associates
Designer: Wladimir Marnich & Anna Sodupe
Client: Rough Luxe Hotel

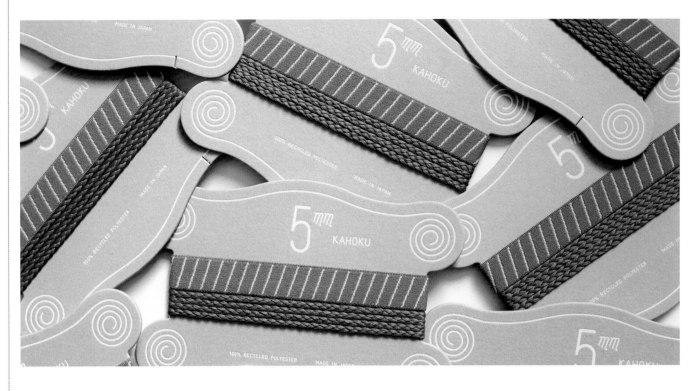

It is designed for an exhibition of textile.

5mm Kahoku Novelty

Design Agency: KOTOHOGI DESIGN

Designer: Naoki Ikegami

Client: 5mm Kahoku

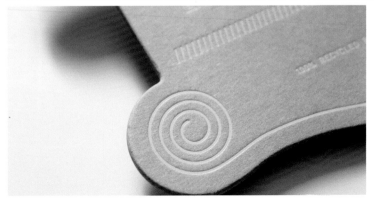

It is designed for an exhibition of paper.

The Silver Ratio Poster

Design Agency: KOTOHOGI DESIGN
Designer: Naoki Ikegami
Client: Heiwa Paper Co., Ltd.

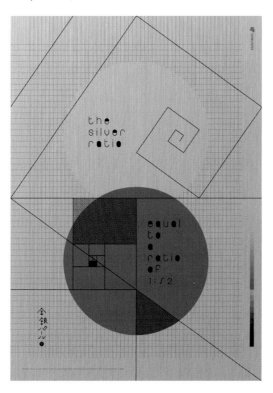
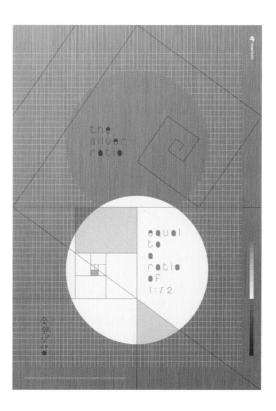
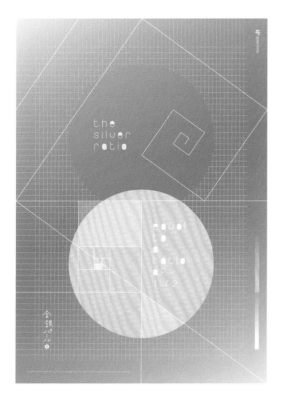
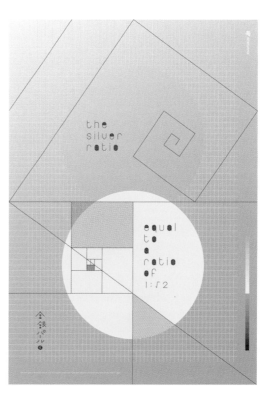

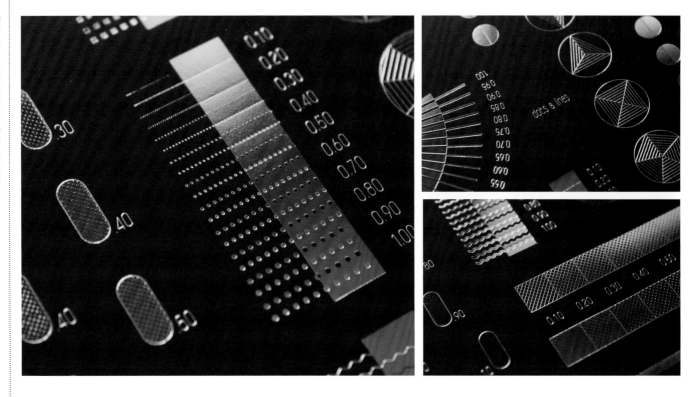

It is a sample of hot stamping.

Dots & Lines Poster

Design Agency: KOTOHOGI DESIGN

Designer: Naoki Ikegami

Client: Cosmotech Inc.

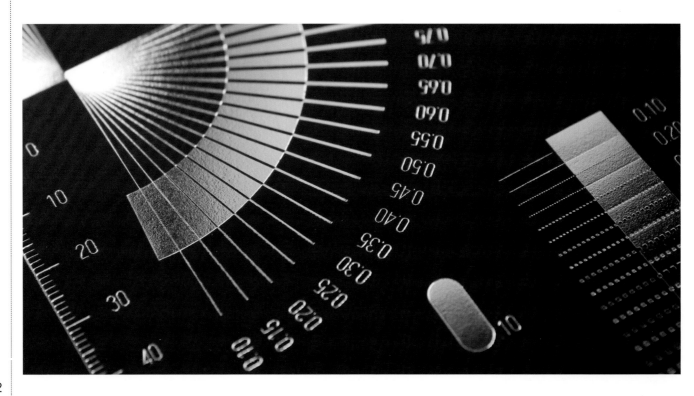

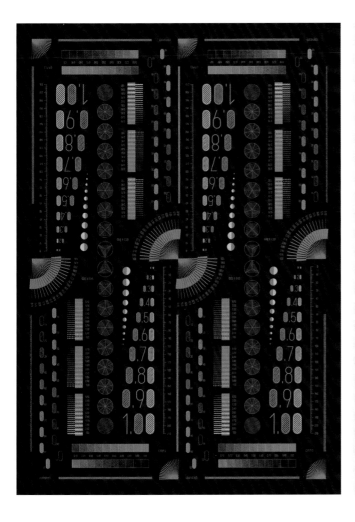
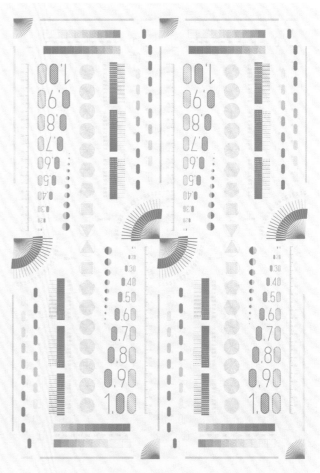

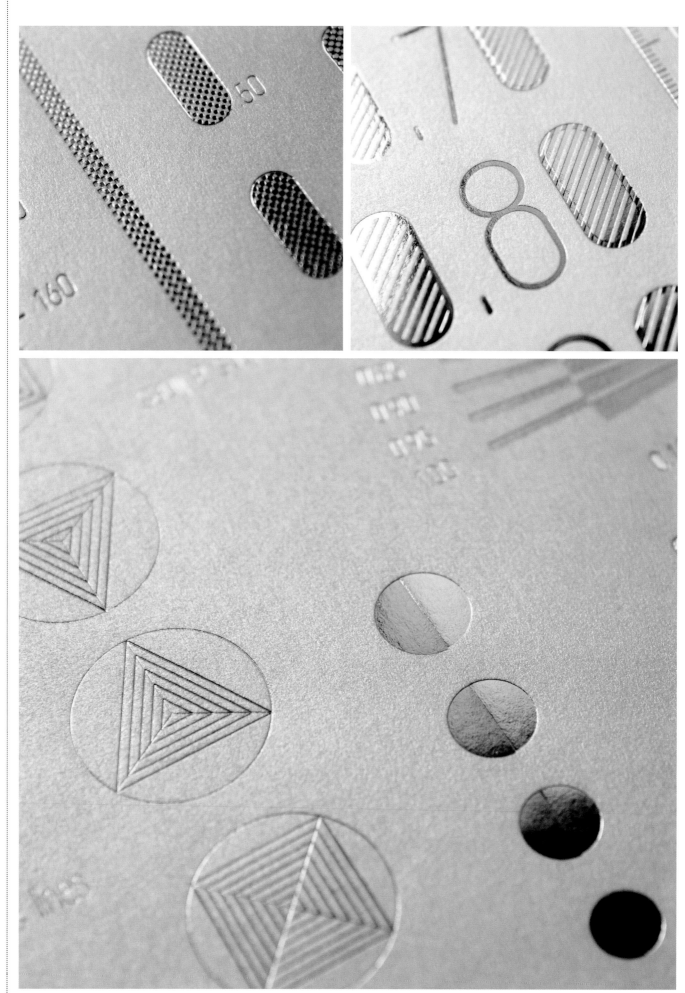

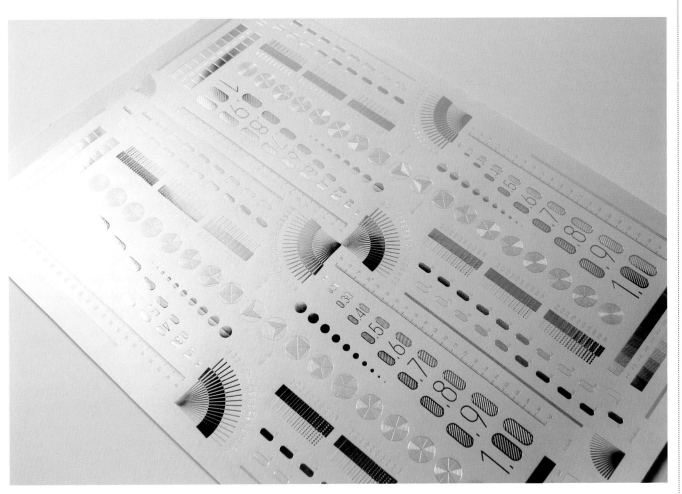

It is a visual identity for a restaurant.

Kushiage Restaurant "Isa" VI

Design Agency: KOTOHOGI DESIGN
Designer: Naoki Ikegami
Client: Kushiage Restaurant "Isa"

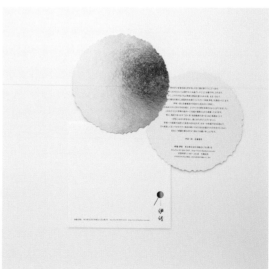

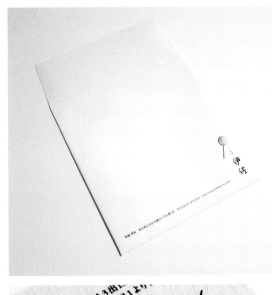
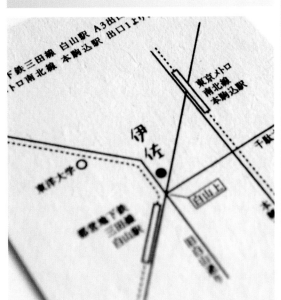
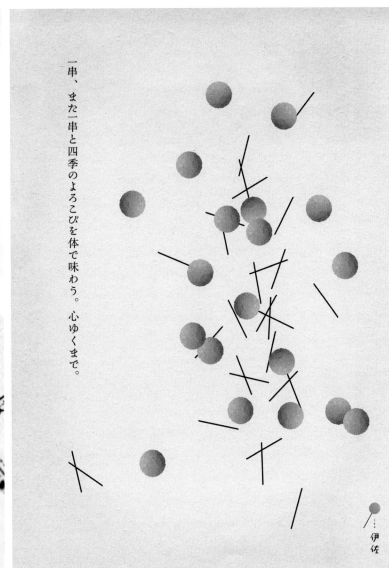
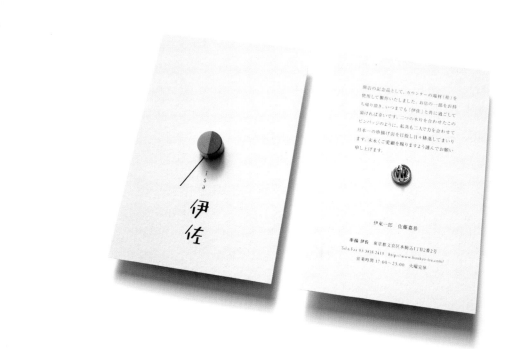

Printing technology: bookbinding, cloths, paper, marveled paper

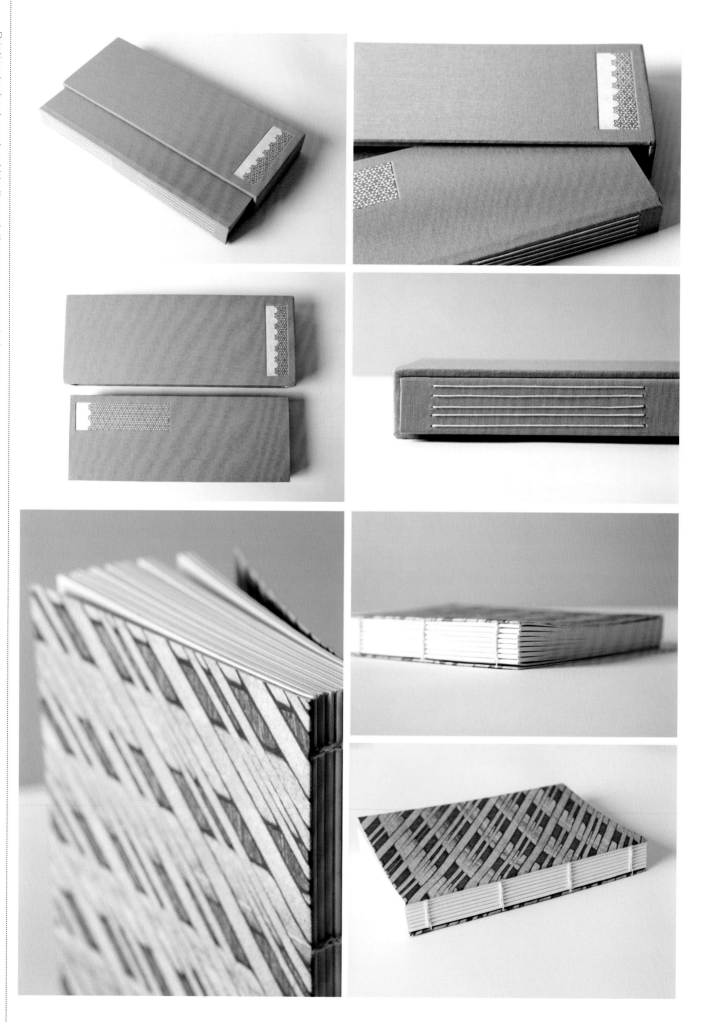

178

It is a collection of handmade books, using the traditional way of binding. Designer loves the smell of the pages, the feel of a book in his hands. He was curious to know how to create a book, how to bind it, and what is the process that happens from the moment. He has got the materials to the final product. That's why he has studied bookbinding.

Bookbinding

Designer: Reut Ashkenazy

Photography: Aviad Bar-Ness

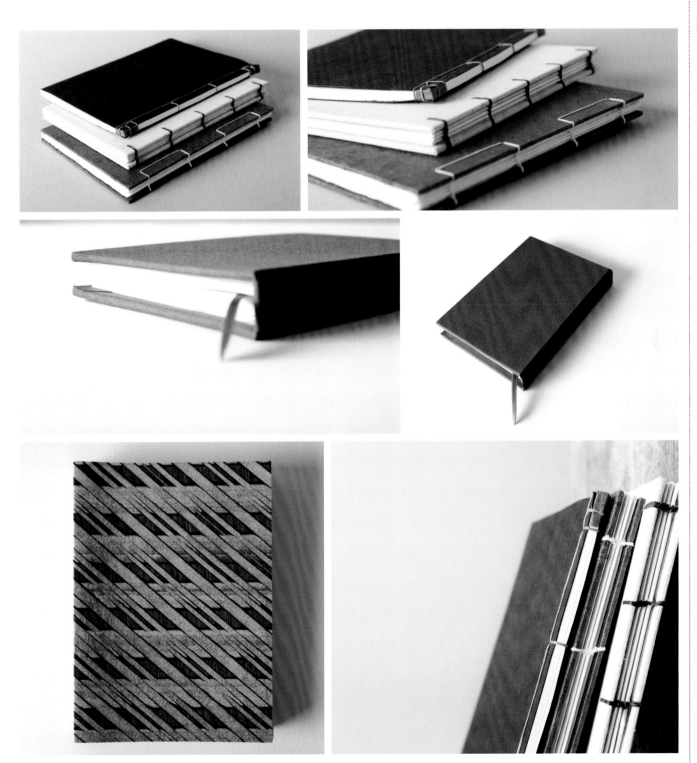

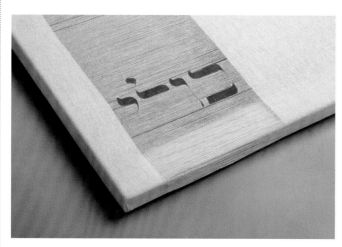
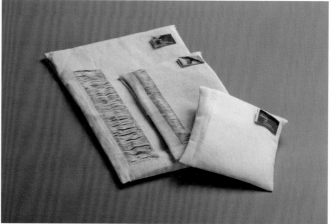

This is a visual identity for cloths' boutique. Kuli (Whole of me - in Hebrew) is a Fashion boutique, which talks about making a whole new person again - spiritually and physically. The boutique specialized in making cloth for disabled people. The concept was based on the Japanese philosophy of the Wabi Sabi, in which the aesthetic is sometimes described as one beauty that is "imperfect, impermanent, and incomplete", and in that sense whole. This philosophy was implemented in the whole project by the disintegration of the paper, different natural fabrics and the images of imperfect nature.

Kuli

Designer: Reut Ashkenazy

Photography: Michael Topyol

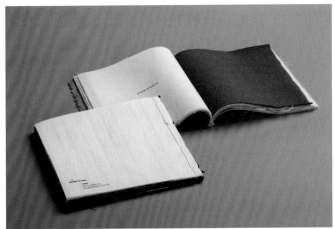

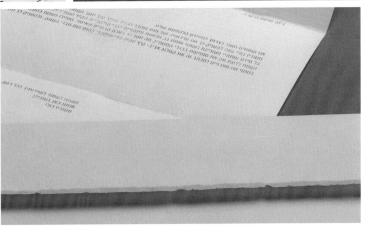

This is an alcohol packaging for women. It is based on the phrase, "in vino veritas" (in wine there is the truth). Much like in this project the phrase suggests that tend to reveal his true feelings under the influence of alcohol.

Normally alcohol companies and manufactures target men as their primer customer. The idea behind this project is to bring a woman's touch and softness to the bottles without losing its presents.

The different fabrics represent different women's characteristics such as cheerful, homey, dominatrix etc., and these are equivalent to the different alcohol drinks such as Tequila, Rum, Vodka etc., overall there are 6 different alcohol bottles.

The techniques used in manufacturing these bottles are flock print for the labels (that are made from chiffon fabric), knitted and woven fabrics for the bottles' exterior. The bottle itself is made from glass. Printing language: Hebrew.

Layers

Designer: Reut Ashkenazy

Photography: Michael Topyol

The starting idea was a paper object designed by Happycentro; its shape, expression of perpetual precision and pureness of the origami world met studio's intimate passion for special printing techniques to match client's need.

That's why they designed and produced an origami whose folded aspect, in the end, showed the Louis Vuitton logo obtained by adding, layer on layer 8 different finishings. After a subtle pantone grey offset print, they applied 4 different foils and relief, silkscreen pearly print and diecut folding line and trim on an 85 gsm sheet.

Louis Vuitton Invitation Origami

Design Agency: Happycentro

Designer: Federico Galvani, Andrea Manzati

Client: Louis Vuitton

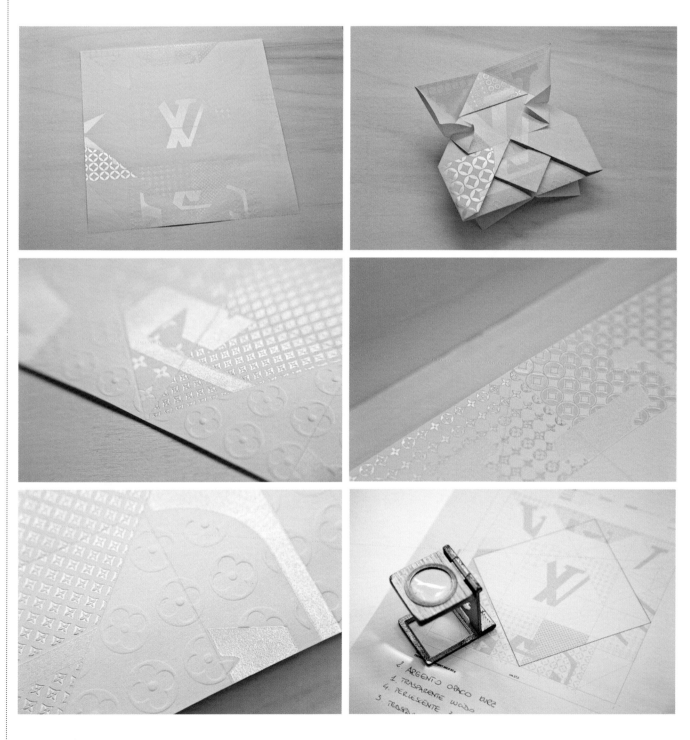

Holger Lippmann, "packing hacking" work series, bas reliefs, 800x600x20mm, 2012.

These works are realized with algorithmic programming using a packing algorithm with processing. Later on, 4 to 5 layers of 4mm painted plywood are laser cutted in different depth (different layers).

For each layer of the laser cutting process, another packing structure is used, so that the overlayering results in a kind of distortions. Sometimes the neet circles are cutted into little pieces again and fall off partly, so that a rather organic form structure comes up.

Packing Hacking Series

Designer: Holger Lippmann

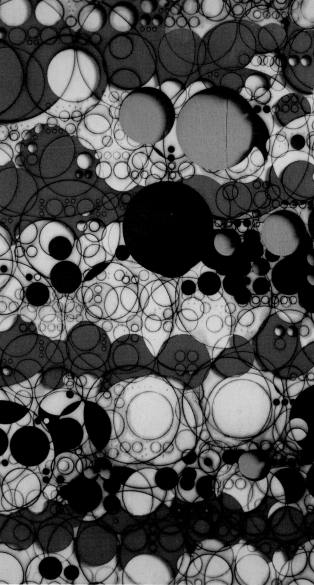

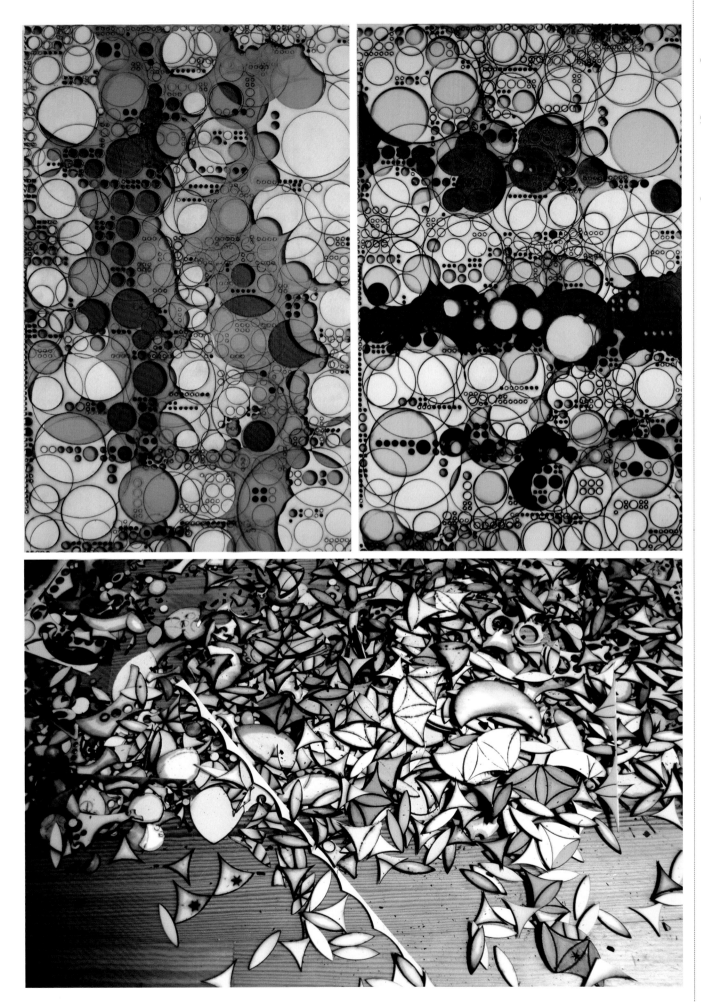

The project was complicated by the fact that this required a very precise register. It was not just "hit a deer in the eye", but the designers managed. For information: the size of the element of cross 0. 43 mm.

Cliche Christmas Card

Designer: Ivan Gulkov

Client: Cliche Letterpress Studio

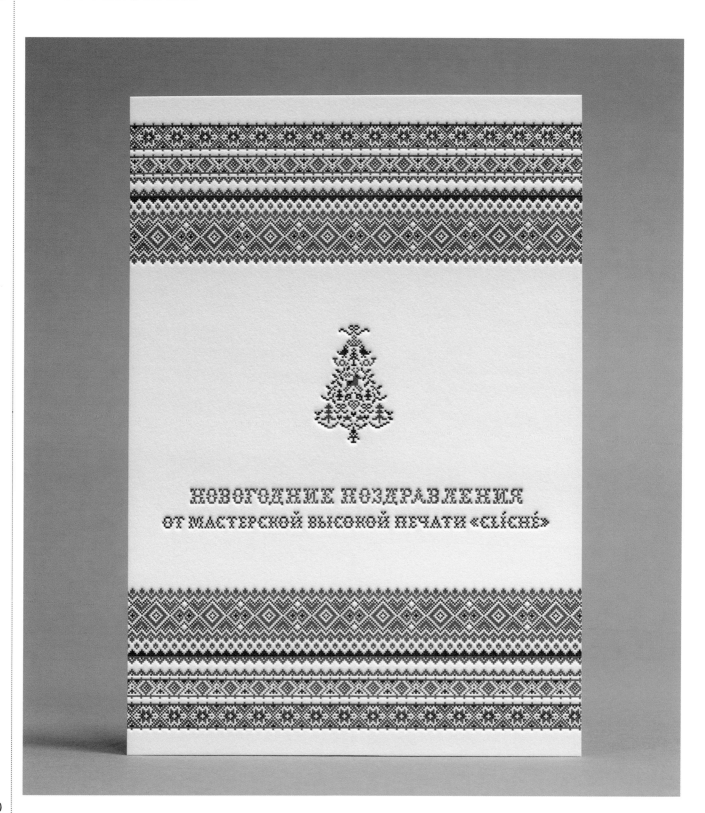

It is a series of culinary posters which present recipes as funny illustration stories.

Culinary Letterpress Posters

Designer: Dima Jo
Client: Demon Press

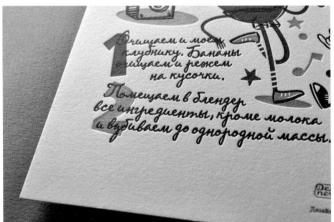

All work was done in Cliche Letterpress Studio.

Greeting Card — MAKE LOVE NOT WORK

Designer: Leonid Zarubin

Client: Lemur Cards

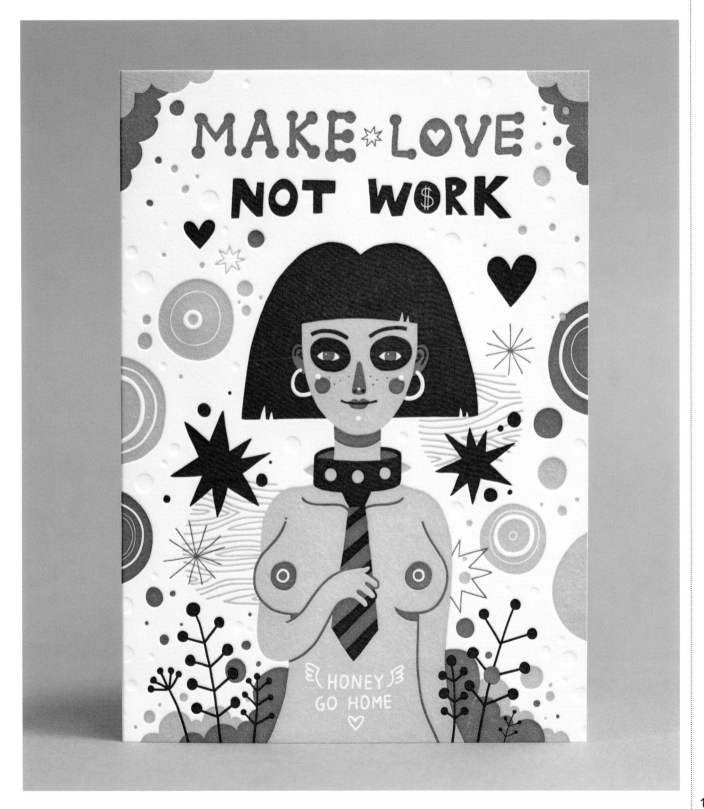

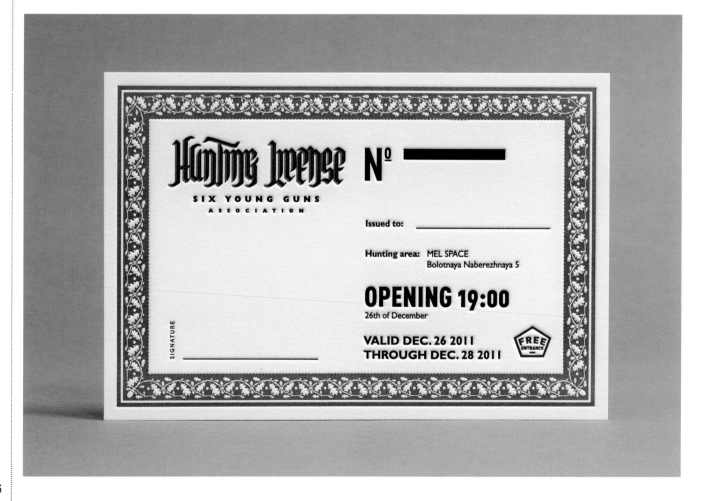

All work was done in Cliche Letterpress Studio.

Invitation — HUNTING LICENSE

Design Agency: Open Season Exhibition
Client: Open Season Exhibition

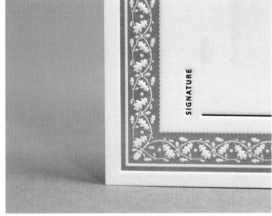

Printing technology: foil stamping, offset printing

This project is self promotional. The designer wanted to send out something that truly was about her and how she design. She didn't want to send out something that was her CV designed in a clever way. The designer wanted to grab someone's attention that had a sense of humour. People take design too seriously, especially when searching for jobs. The designer basically wanted to say "Hey, I know that your busy and have a team full of great people, but I can get the job done too".

Typographic Wank

Designer: Shakira Twigden

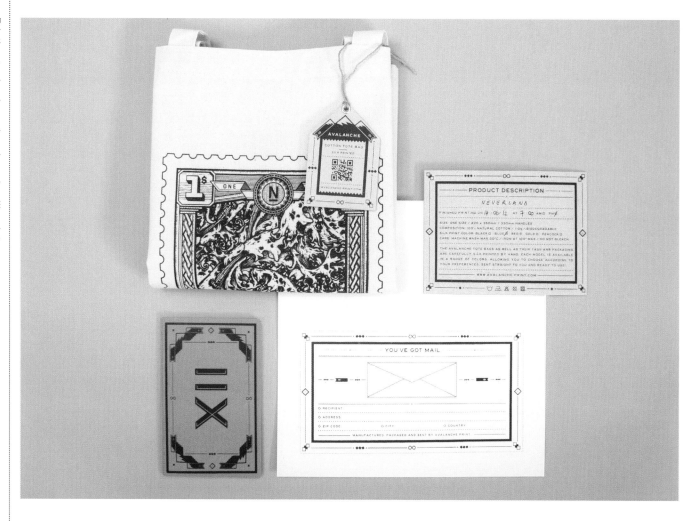

Avalanche Print is an online platform offering a choice of silk printed tote bags as well as silk printed notebooks. There is a range of 7 models and 5 colors to choose from. Every item is silk printed one by one, and the project's whole idea is to highlight the process of hand-made silk print and to make it popular again.

Avalanche Print

Design Agency: Say What Studio

Client: Avalanche Print

Photography: Pierre-Luc Baron-Moreau

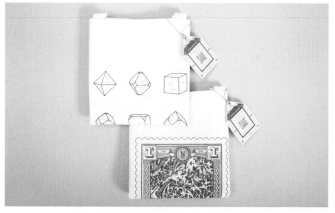
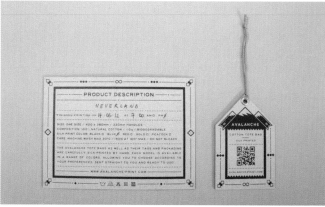

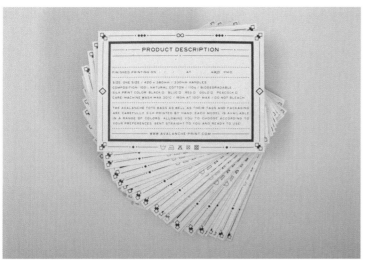

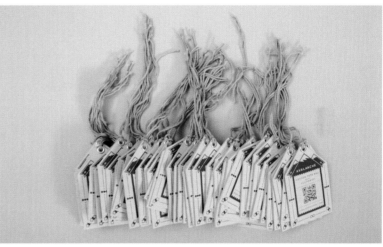

Das Seine. Forschungs Projekt catalogue was designed for Galeria Piekar, Poland.

Das Seine. Forschungs Projekt

Design Agency: 3group
Designer: Ryszard Bienert
Client: Galeria Piekary

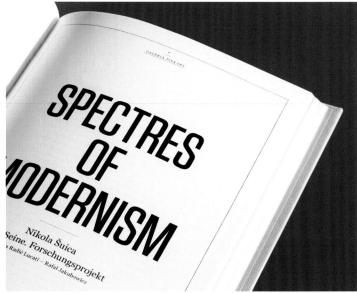

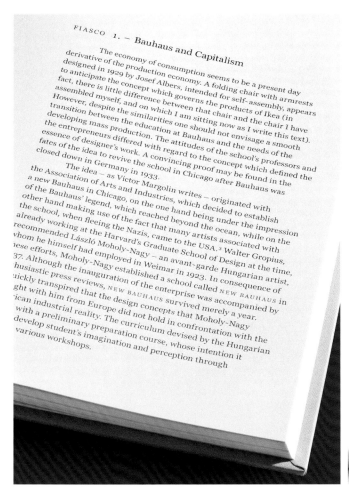
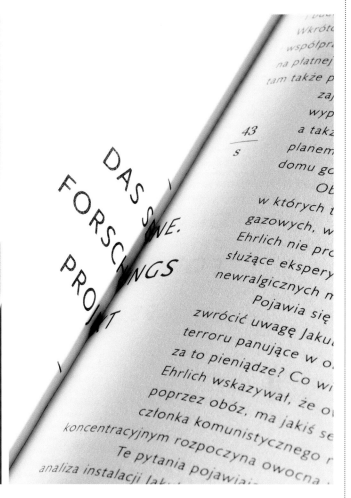

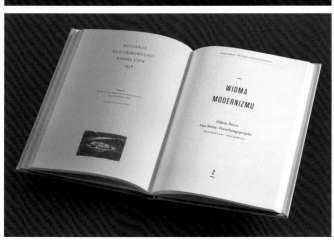

Oh no, not sex and death again! catalogue was designed for Panstwowa Galeria w Sopocie, Poland.

Oh No, Not Sex and Death Again!

Design Agency: 3group
Designer: Ryszard Bienert
Client: Panstwowa Galeria w Sopocie

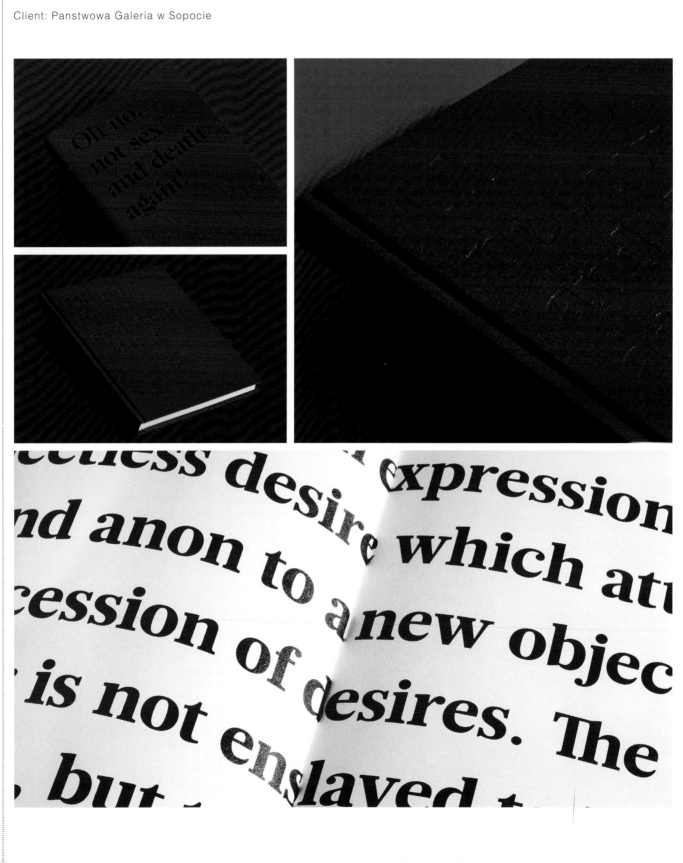

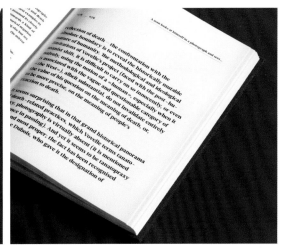

Printing technology: foil application, blind debossing, bookbinding and offset printing

Julien Hauchecorne is an art and antique merchant based in Paris, specialized in vintage 70's furniture design and art.

Business cards are printed on 2 subtle duplexed uncoated smooth paper manufactured by GFSmith to reach 700 gsm and both cards have a deep debossed monogram logo on the front side. The 500 cards are printed on the backside with a diffraction effect foil on a pristine white and ebony black duplexed substrate with a diffraction foil fore-edge printing process on the outside edges.

Julien Hauchecorne Business Card | Diffraction Version

Design Agency: DMWORKROOM

Designer: Denis Mallet

Client: Julien Hauchecorne

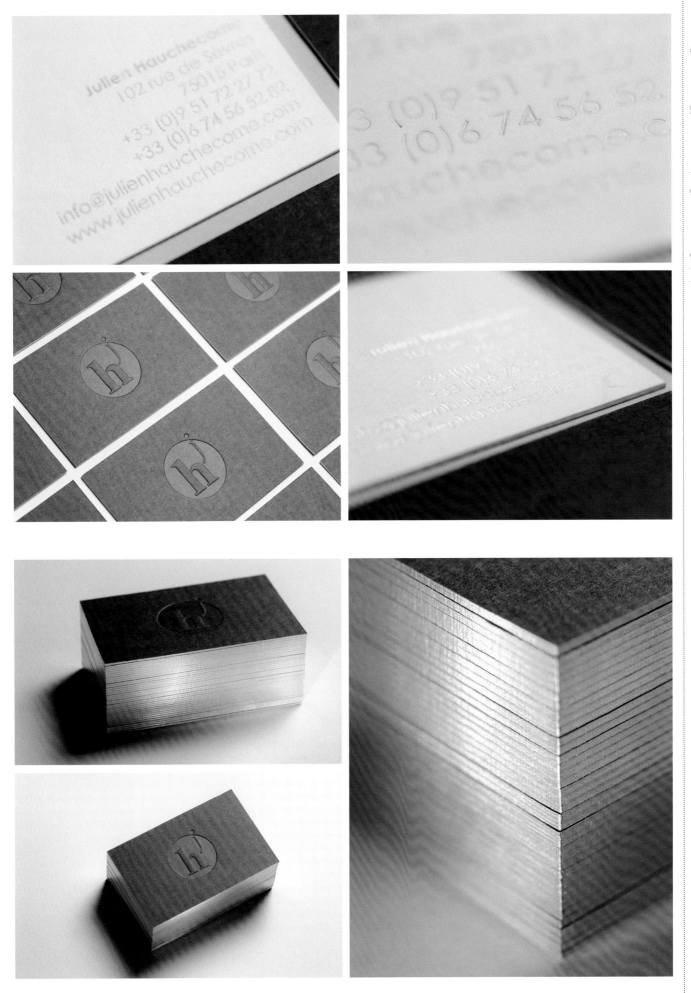

Julien Hauchecorne is an art and antique merchant based in Paris, specialized in vintage 70's furniture design and art.

Business cards are printed on two subtle duplexed uncoated smooth paper manufactured by GFSmith to reach 700 gsm and both cards have a deep debossed monogram logo on the front side. The 500 cards are printed on the backside with a powerful lime green foil on ebony black and dark grey duplexed substrate with a lime green foil fore-edge printing process on the outside edges of the business cards.

Julien Hauchecorne Business Card | Lime-green Version

Design Agency: DMWORKROOM

Designer: Denis Mallet

Client: Julien Hauchecorne

The designer designed the visual identity around a geometrical pattern that was creating network connectivity. Placing the letters in the center of each of the graphic polygonal shapes allowed the name to be the core of the geometrical network pattern. Extracting the letter "M" and "T" with their surrounding polygonal structures created a logotype working as an ingenuous monogram. The simplicity of this geometrical shape allows the use of many different high finish print processes, as metal and black foil blocking on the compliment slips and business card, blind embossing on the envelops and photo chemical etching on the metal side of the business card. The monogram can also be easily stamped using an embosser or rubber stamping machine to personalize letters and cards, and for authenticating important documents.

Business card paper side was printed on a 350gsm Black Ebony subtle uncoated smooth paper with new leather embossed surface manufactured by GF Smith. The business card metal side was chemically etched on a high-grade 0.4mm stainless steel metal board.

Compliment slips are manufactured with duplexed uncoated smooth boards produced by GF Smith. The front side is stamped with a glossy black foil on a 350gsm Black Ebony paper substrate and the backside is stamped with the same foil on a 350gsm Pristine White paper substrate, making the finished slips a 700gsm board.
Mathias Tanguy identity is set up using Arno Pro typeface.

Mathias Tanguy Brand Visual Identity

Design Agency: DMWORKROOM

Designer: Denis Mallet

Client: Mathias Tanguy

This poster depicts a regular housefly in high resolution detail. The whole illustration consists of many Levi's Buttons, thus it becomes the original Button Fly, obviously.

The poster is printed with special printing techniques: the background is dark silver and the button graphics are foil stamped with an eye catching and extravagant rainbow foil.

Levi's, Button Fly

Design Agency: Sagmeister & Walsh

Deisgner: Richard The, Joe Shouldice

Client: Levi Strauss & Co

This is the stationery design for N. Daniels, a rep and photo producer in Vienna. It's simple, cool and thermo sensitive. The black color of the varnish fades at body temperature as soon as you hold it in your hands, you literally produce an image by yourself. It's a dynamic and living design — the business cards are little polaroids with a constantly changing surface. All these cards might start looking similar, but with your personal "touch", you brand them individually.

N. Daniels Wien

Design Agency: BUREAU RABENSTEINER

Designer: Isabella Meischberger

Client: N. Daniels

Photography: Natalie Daniels

Printing technology: silk screen printing with thermo sensitive varnish

Printing technology: silk screen printing with thermo sensitive varnish

Printing technology: silk screen printing with thermo sensitive varnish

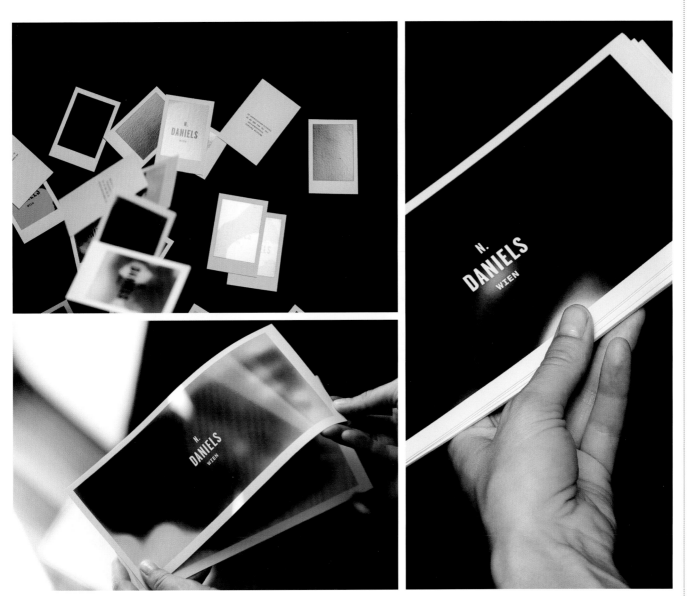

Printing technology: silk screen printing with thermo sensitive varnish

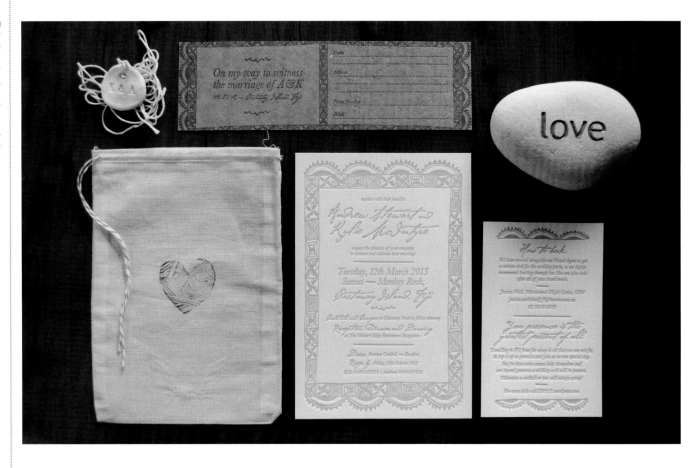

Wedding invitation set designed to reflect the location, and theme for the destination wedding. It is inspired by Fiji Tapa Art and colour theme for the wedding. Letterpress provided a touch of elegance to the invitation.

Andrew & Kylie — Fiji Wedding Invitation

Design Agency: Peachy Flamingo

Designer: Kylie McIntyre

Client: Kylie & Andrew

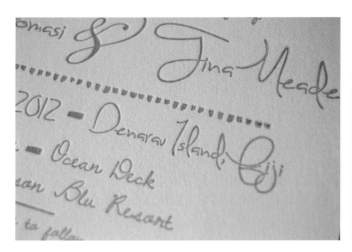

Wedding Invitation set designed to reflect the location, and theme for the destination wedding. It is inspired by Fiji Tapa Art and colour theme of the wedding. Hand Illustrations provided the rustic feel to the invitation and letterpress added a touch of romance.

Fiji Wedding Invitation

Design Agency: Peachy Flamingo

Designer: Kylie McIntyre

Client: Tina & Jale

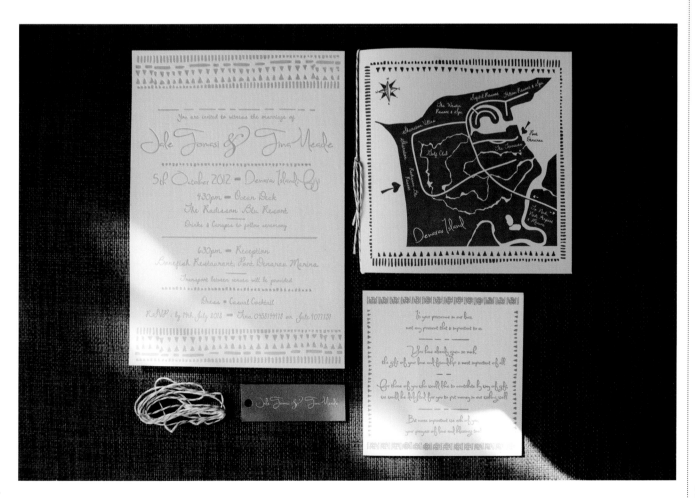

Identity and stationery for an architecture and interior practice based in East London. The design is inspired by architectural hatch patterns which are used to distinguish different building materials. Different hatch patterns are used on the back of the business card and in small strips as part of the logo.

Blustin Design Stationery

Design Agency: Mind Design

Client: Blustin Design

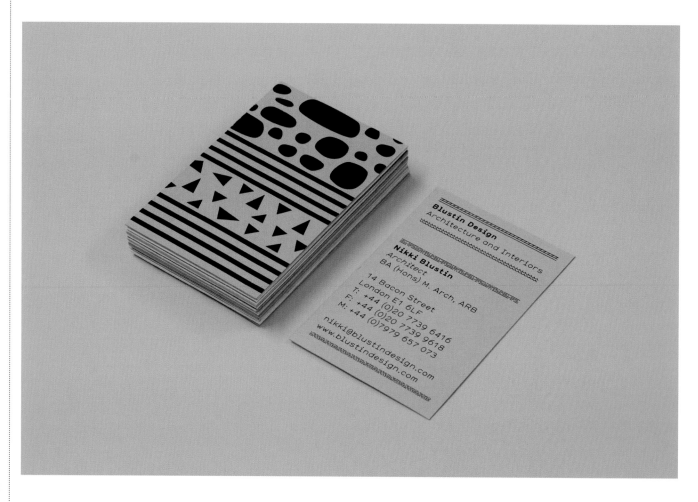

Printing technology: blind embossing, rubber stamping

Identity and stationery for Playlab, a workshop space aimed to be a creative playground for stressed adults. The illustrations used are a mixture of scientific elements and random funny images. The stationery is printed in fluorescent Pantone colours while the actual logo is just blind embossed. The address details are filled in using a rubber stamp.

Playlab Identity

Design Agency: Mind Design

Client: Playlab

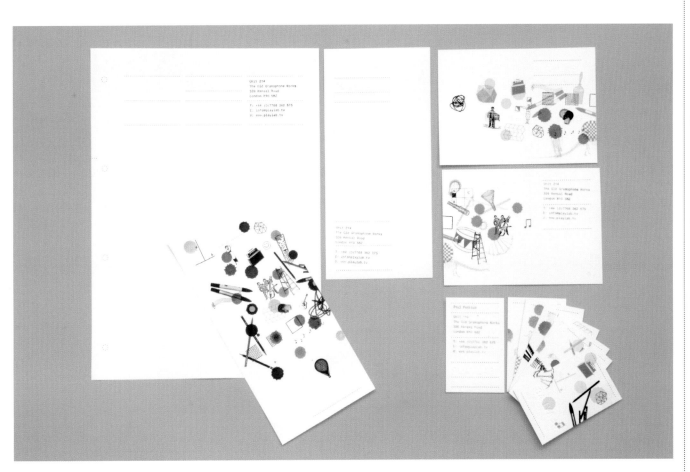

Printing technology: blind embossing, rubber stamping

Printing technology: blind embossing, rubber stamping

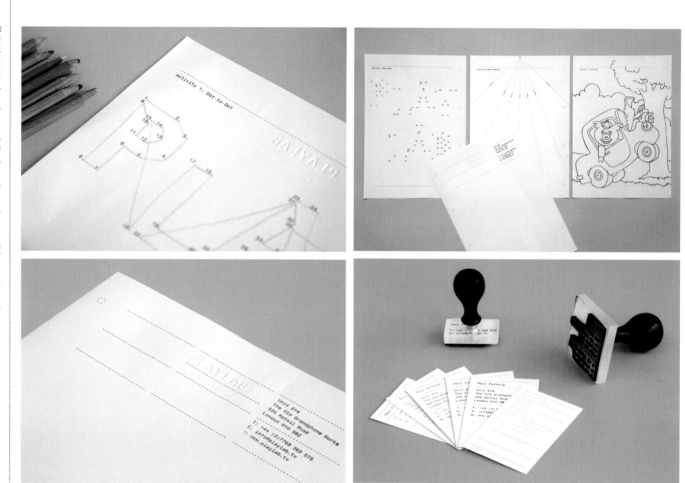

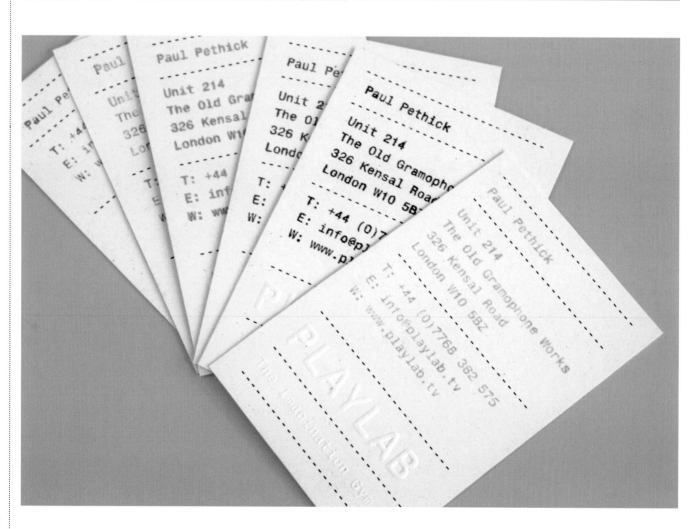

Printing technology: blind embossing, rubber stamping

227

The Whitest Boy Alive's Long Play is a personal project which born motivated by the desire of doing some work using the foil stamping technique. The diagram chosen to highlight this technique is the contrast between very thin strokes and plain black. The Long Play's cover was print with gold and black ink, and CMYK inside.

Long Play for TWBA

Design Agency: Andres Reisinger

Designer: Andres Reisinger

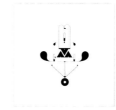

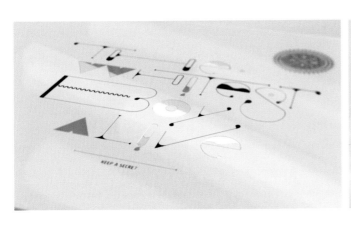
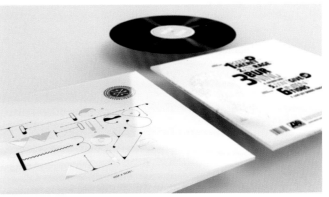
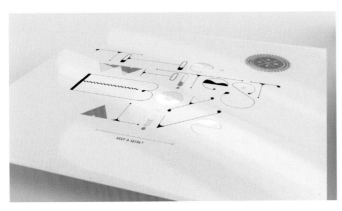

Printing technology: intaglio printing, photopolymer plate, etching press, ink wash

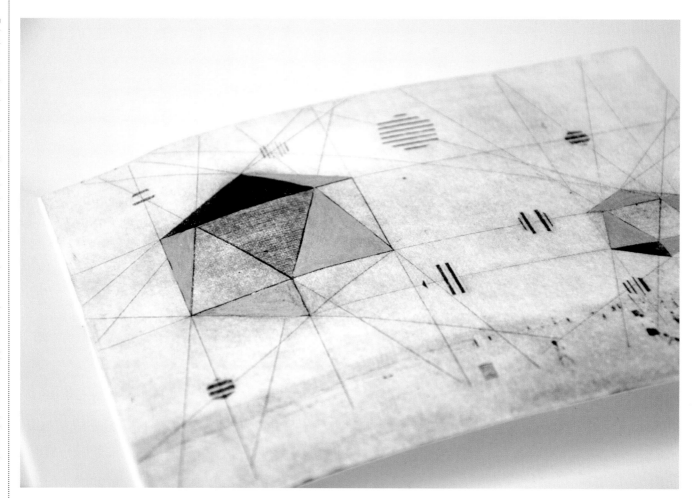

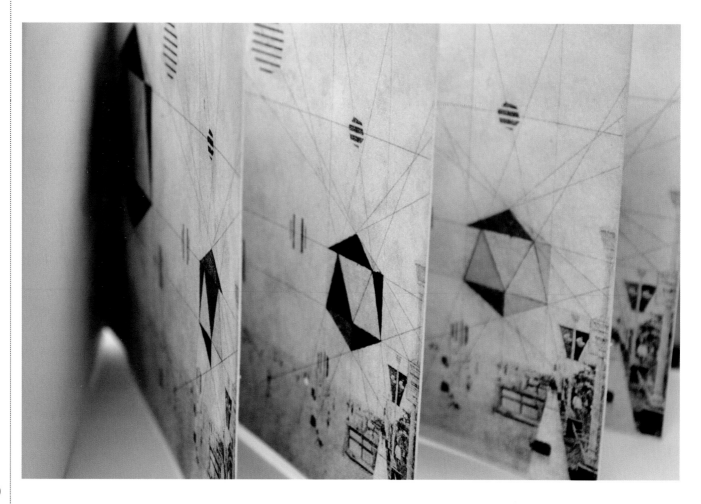

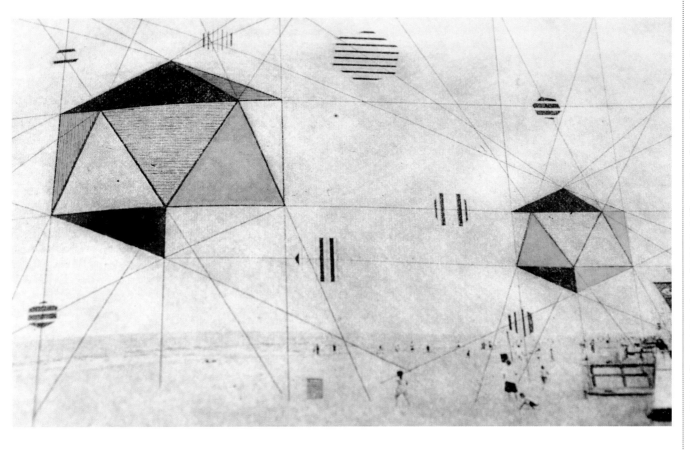

Icosahedrons is a set of prints based on the relationship between platonic solids, geometry, and the beauty of nature. Combining digital vector artwork and photography, the design was transferred to a photopolymer plate for the intaglio print process. Each print had a unique colored ink wash applied to it after printing.

Icosahedrons

Designer: Annalisa Kleinschmidt

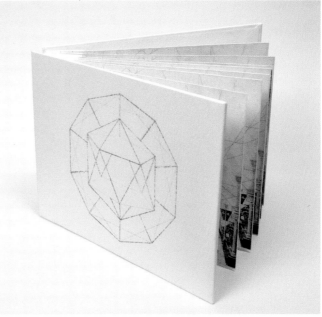

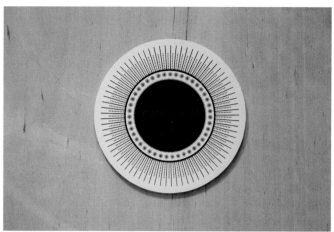

Seamless Creative wanted to design a celebratory and useful client promotion to ring in the New Year. Once they decided that coasters fit the bill, it was a no brainer to adorn them with typographic exclamations of "Cheers!". It was an expression that would work for all sorts of occasions, all year round. The designers stuck with a somewhat neutral but sophisticated, palette of navy and metallic gold to reinforce the festive theme.

2012 Cheers Coasters

Design Agency: Seamless Creative

Designer: Courtney Eliseo, Brian Eliseo

Client: Self-promotion

Contributor: Mama's Sauce

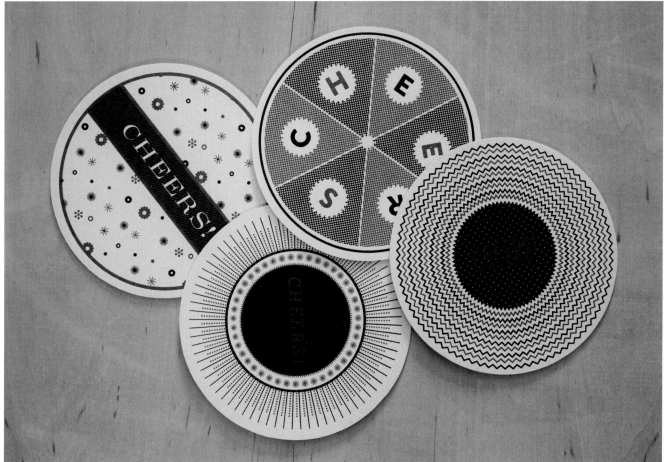

Printing technology: screen printing

The culture of the Cavalli Stud & Wine farm located in the heart of the Stellenbosch winelands is deeply rooted in a passion for horses.

This passion has been made tangible through Cavalli's identity which reflects the spirit, grace and power (yet unpredictability) of the animal in every aspect of its manifestation.

Cavalli Corporate Identity

Design Agency: Studio Botes
Designer: Brandt Botes
Client: Cavalli Stud & Wine Farm

Printing technology: offset litho, gold & black foil stamping, die-cutting, spot gloss UV varnish (folders inside)

Printing technology: offset litho, gold & black foil stamping, die-cutting, spot gloss UV varnish (folders inside)

Printing technology: offset litho, gold & black foil stamping, die-cutting, spot gloss UV varnish (folders inside)

Printing technology: offset litho, gold & black foil stamping, die-cutting, spot gloss UV varnish (folders inside)

This has been sprayed with glass powder on the special ink that is silk-screen printing, glass powder reflects light.

Company Brochure "Entrance"

Design Agency: KucHen
Designer: Etsuko Iwasaki

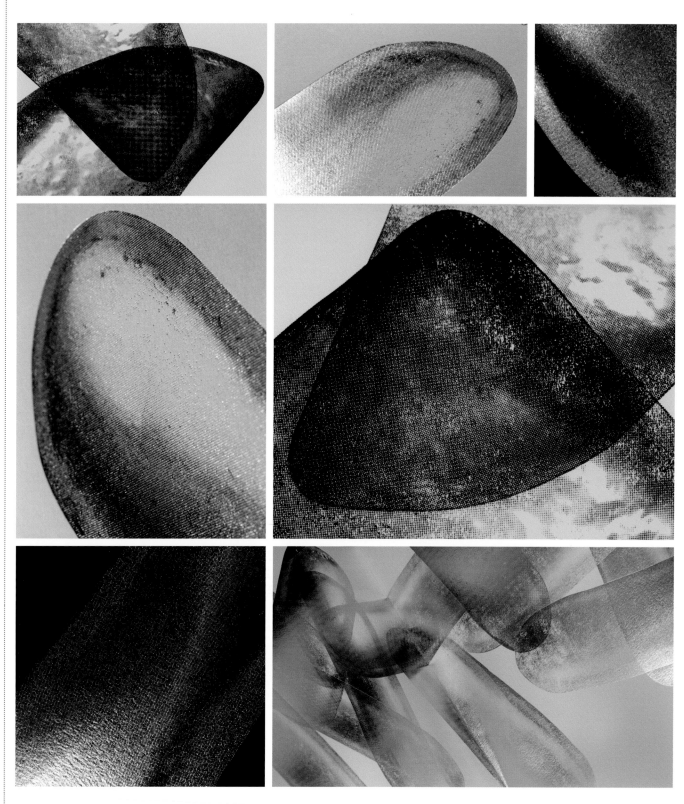

This is the art book made in experimental techniques such as to express gradation of depth in the gold foil.

E. Company Brochure 2012 Foil Book

Design Agency: KucHen
Designer: Nahoko Kusumi

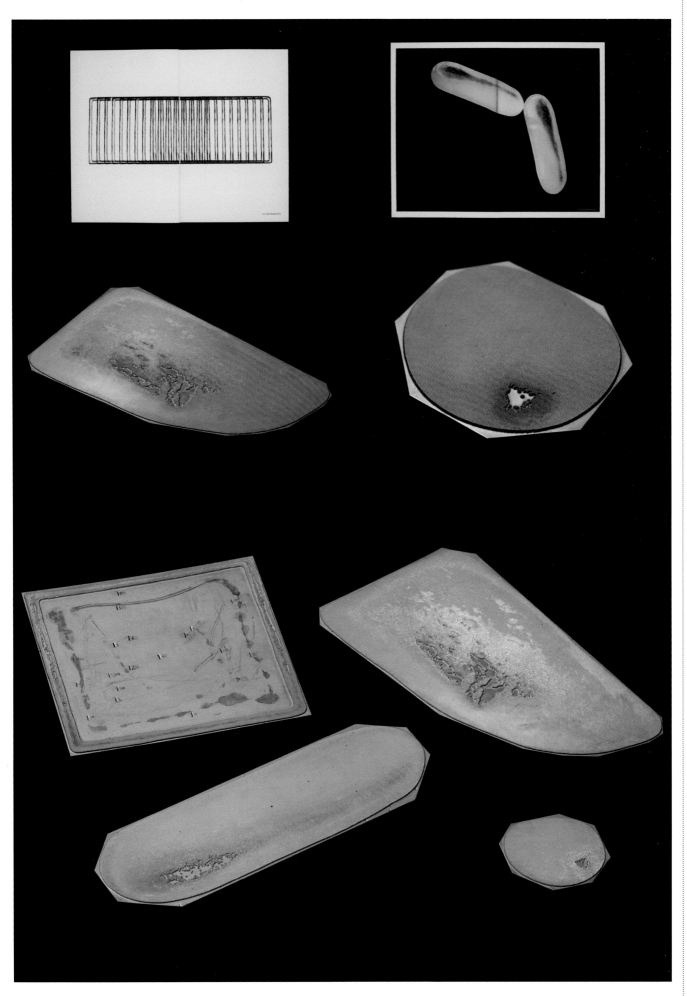

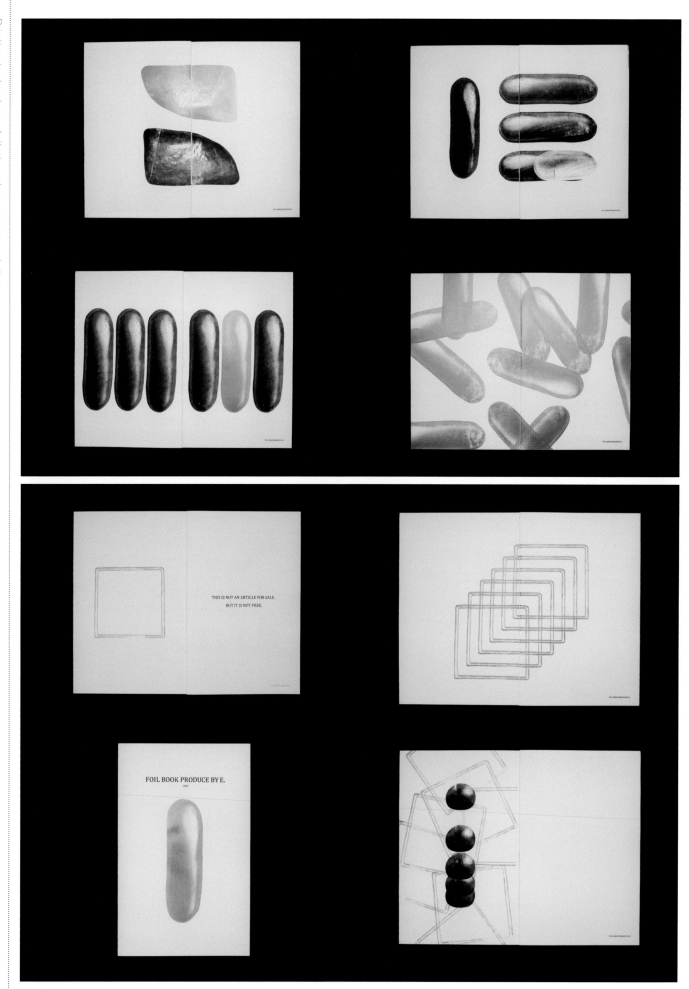

It is a business card for Green Sky Media by Hear Agency. This card is printed on 16pt stock featuring a silk coating, spot UV, rounded corners and a great modern design.

Green Sky Media Business Card

Design Agency: Hear Agency

Designer: Chase Houston

Client: Green Sky Media

Business card designed internally for Hear Agency. rounded corners, spot UV, silk lamination, and 16pt stock all make this unique and well designed business card stand out!

Hear Agency Business Card

Design Agency: Hear Agency

Designer: Chase Houston

Client: Internal for Hear Agency

Founded in 2007, a New York based women's fashion brand, IDEEËN, combines structural shapes and soft draping techniques with a New York street wear vibe. The theme of IDEEËN Autumn Winter 2009 collection was "deep sea". STUDIO NEWWORK was assigned to create the presentation invitation that implied the theme. The timetable and other information were placed as if they were floating. Inspired by deep-sea fish skin, STUDIO NEWWORK chose to print on silver metallic paper.

IDEEËN Autumn Winter Invitation

Design Agency: STUDIO NEWWORK

Client: IDEEËN

"This poster is a typographic project made out of Bodoni letters being inspired by the female sensitivity. Bodoni is a series of serif typefaces first designed by Giambattista Bodoni (1740 - 1813) in 1798. Like every typeface, Bodoni has its own charisma. Its presence in design always reflected the good taste, the classic, the elegant, and the different. Bodoni has narrower underlying structure with flat and unbracketed serifs, extreme contrast between thick & thin strokes and an overall geometric construction.

The reason designer decided to use silkscreen as a method was to add a little spark using gold colours and gold foils to reflect the ultimate woman nobility."

Bodoni Girl

Designer: Andreas Xenoulis

Client: Andreas Xenoulis

Printmaker: tind

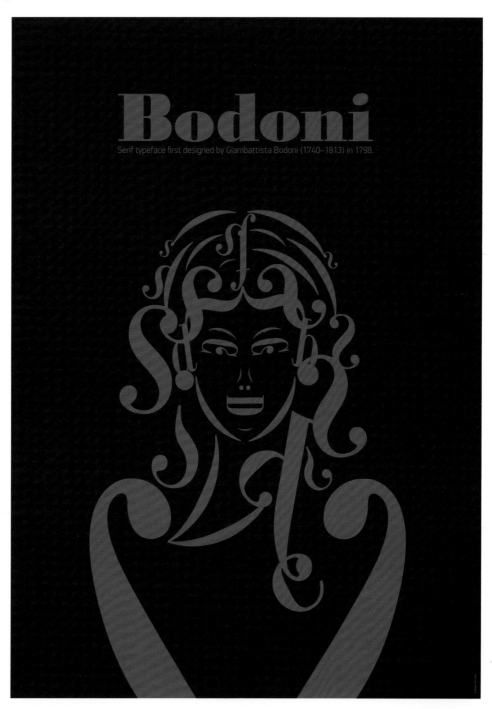

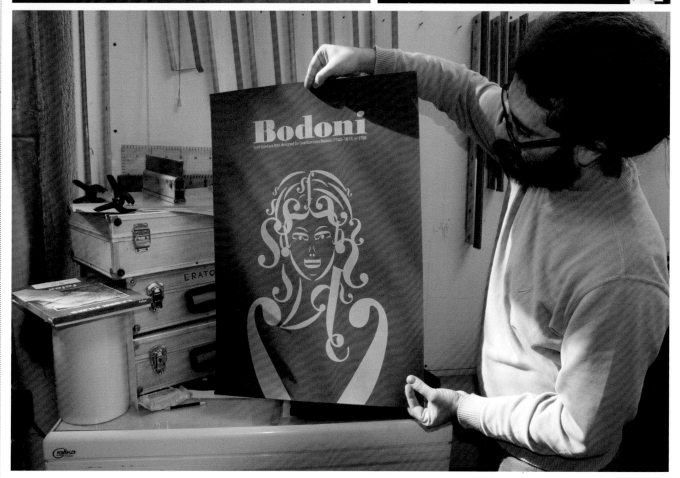

It is about future & past, fluorescent & black. There are 2 layers of the hand printed poster.

Designed with custom designed macro tools on Visual Basic for Corel Draw by Erato. The poster is an attempt to make their own raster. It is silkscreen hand printed with fluorescent colors and the blackest black by tind.

The Future is in the Past

Design Agency: tind

Printmaker: tind

RockShox is a company owned by SRAM which produces front and rear shocks for bicycles. The concept and format of the catalogue based on their tagline "The Earth Is Not Flat". So die-cut pages of the catalogue represent the contour lines of the topographic maps and the cover is also printed with sculpture embossing technique to reflect the terrain location that matches the die-cutting.

RockShox Catalogue

Design Agency: TNOP™ DESIGN

Designer: Tnop Wangsillapakun

Client: RockShox, and Segura Inc., Chicago

The calendar should faithfully represent the brand philosophy and basic values. HUGGIES brand philosophy implies joy, pleasure, freedom of discovery and learning of the wonderful world, comfort and liberty of welfare. HUGGIES allows a baby to walk easily through the life, learn the surrounding world and not to care of the "small problems" that might happen. In HUGGIES' eyes, each baby is an individual. The child's world is unique and amazing and play is a serious business.

The designers have created the idea of two "baby's feet", that move along the page imitating the baby's steps and showing the current day. Every new day is a step forward. The surrounding world is shown as a pattern of simple things from which a baby starts to discover the world.

Walking Calendar

Designer: Yurko Gutsulyak

Client: Huggies, Kimberly-Clark Ukraine

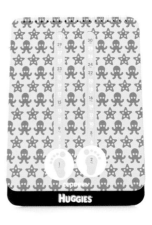

This is the 300 page bid submission book project to make Abu Dhabi the headquarters of the International Renewable Energy Agency. This was a collaboration project with GMMB, an ad-agency from Washington DC. The concept of the book is to show the unity of the sources of renewable energy in the region and the city of Abu Dhabi and the UAE as a whole. The design elements on the book case and the cover also inspired by the traditional white dish dash or the long white robes that the men in that region wear, the impressions on the sand, the orange color sunset and the Arabesque patterns in their architecture. The special printing techniques, including laser die-cutting, blind debossing, foil stamping and laminate that help enhancing the design of the book.

UAE's Bid Book

Design Agency: TNOP™ DESIGN

Designer: Tnop Wangsillapakun

Client: GMMB, D.C., Masdar, the U.A.E.

Printing technology: foil stamping, blind debossing, laser die-cutting, laminating, hand-converted slipcase

A personalized notebook for designers has to be inspiring. That's why the mix between a didot type, delicate ornaments and light grid match perfectly with laser engraving.

The color purple suggests wealth and quality. With the silver metallic Pantone, the graphics on the card fit nicely with the notebook.

Personalized Notebook + New Year Card

Design Agency: ReflexParis

Designer: Florent Carlier

Printing technology: laser engraving, offset printing

Oodles of ink, textured materials and the freedom of printing without a machine are the identity of traditional screen-printing. Being an illustrator, Ruchi was naturally drawn to this process. After spending 6 months messing about with thick inks, mixing the wrong chemicals and tearing a few screens in her small family-run press in India, Ruchi conceptualized and printed this book to share some successful explorations. Having managed to select most of paper and materials that would become a printer's nightmare, she has converted them into attractive artworks that the viewer simply cannot resist touching.

Screenprinted Handbook

Designer: Ruchi Shah

Printing technology: offset, embossing, debossing, 3 different foils, silver printing, die-cutting, silver laminating, spot UV, screen printing

It is a delivering slick design for tax-free drivers — intelligent packs for clever car scheme.

Autopia provides a unique service in Australia, helping individuals and employers maximize potential tax savings through novated leasing programs.

They needed drivers packs with a special quality that drivers would value, as it is their first touch point with new customers. Early off the shelf prototypes were too disposable, so they turned to Synsation for a bespoke driver guide and welcome pack — as well a broad range of branded collateral.

The pack, which includes fuel credit cards and guidebook, features a distinctive sleeve and folder with 3 color foils, debossing, embossing, silver ink, die-cutting and aluminum laminate. Drivers are encouraged to keep it in their glove box. The packs are being sent out to new leaseholders now, having exceeded Autopia's expectations in every way.

Autopia Driver Guide

Design Agency: Synsation Brand Design

Designer: Sander Dijkstra, Scott Robertson

Client: Autopia

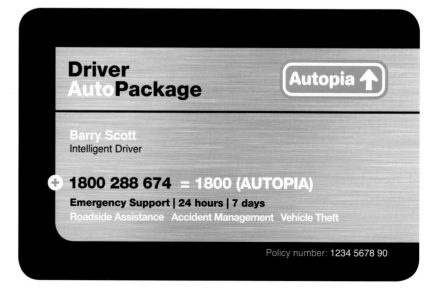

Printing technology: offset, embossing, debossing, 3 different foils, silver printing, die-cutting, silver laminating, spot UV, screen printing

Printing technology: offset, PMS 877 metallic ink, die-cutting and insertion of a pencil into the cover, silver foil guilding

American Odysseys is an anthology of 22 novelists, poets, and short story writers.

The title of this book is printed on a pencil which is then inserted into a diecut in the cover. The pencil was chosen as a symbol of the writer and the writing process. For many of people, the pencil is the first introduction to creating letters, words, and later on sentences. The stories contained within are of a personal nature, and the intimate quality of the pencil seemed to be the perfect complement. The pencil is completely removable.

American Odysseys: Writings by New Americans

Design Agency: yesyesyes design

Designer: Joe Shouldice

Client: The Vilcek Foundation

Everyone loves gifts. This holiday card let's you unveil over 280 unique gifts hidden under scratch-off wrapping.

Vilcek Holiday Card

Design Agency: yesyesyes design
Designer: Joe Shouldice
Client: The Vilcek Foundation

Printing technology: offset, flood coat of release varnish, spot "soft-touch" matte varnish (outside the gifts so that it had a nice velvety feel), 1st screen of scratch off foil (silver bottom), 2nd screen of scratch off foil (the ribbon)

You are important. That's why the designers acknowledge your presence even on their cards. Created with a series of mirrored foils, every facet has a potential to alter a light ray. Your uniqueness is pivotal as designers capture your potential at your brightest. They reflect the radiance in each other and the brilliance sparking from the perfection of their collaboration shines infinite.

You & Me Creative's Business Card

Design Agency: You & Me Creative

Designer: Edwin & Jasmine

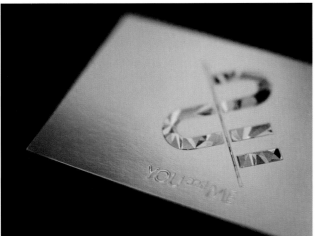

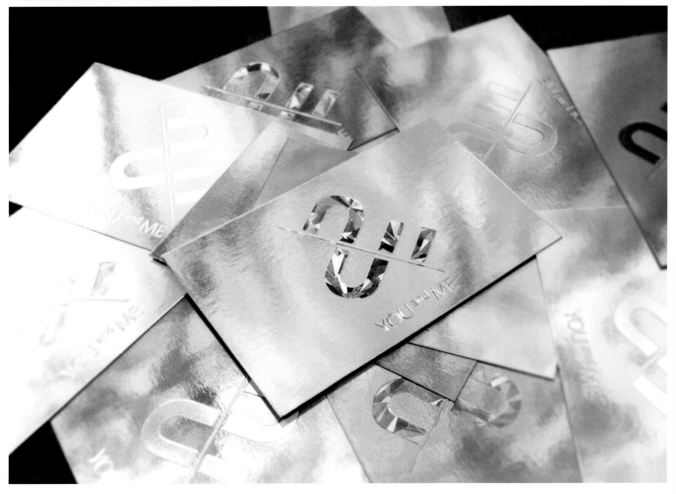

Collaboration between Sylvia Gottwald and Institut Parfumeur Flores saw perfumery walls display some incredible mother of pearl jewels over the Christmas period. The designers created an invitation, which uses holographic foil on a thick black card in order to represent nacre in the closest possible way. Sylvia's designs with pearls & nacre are unique pieces in limited editions, created from 7 species of pearl producing shells and various fresh water pearls.

Pearls & Perfumes

Design Agency: Bunch

Client: Institut Parfumeur Flores

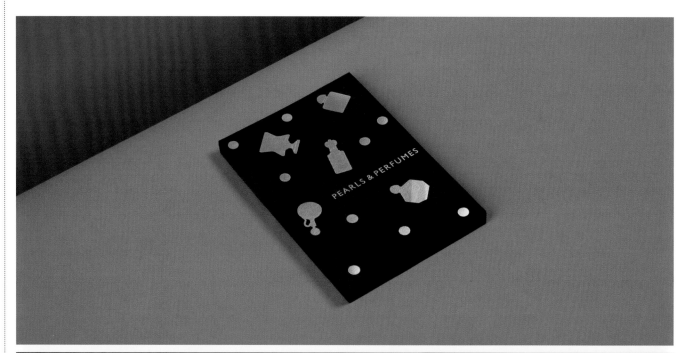

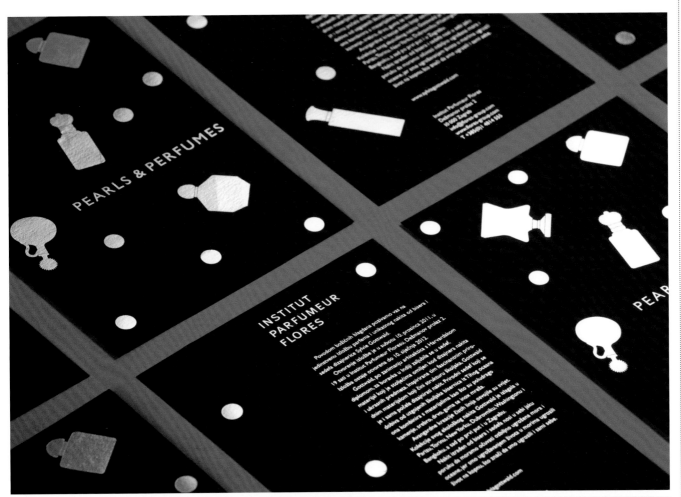
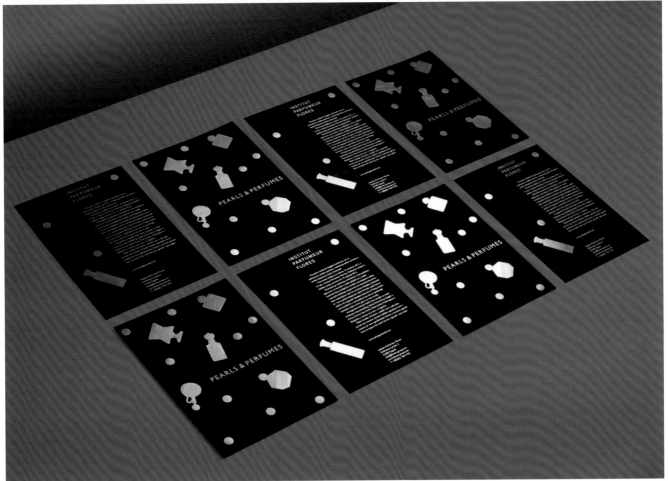

A fresh and particular label. A classic theme becomes modern and fresh with the UV print on the details and the woman's profile diecut out creating a difference in texture and a beautiful transparent effect.

Altadonna

Design Agency: Doni & Associati

Client: Altadonna

Printing technology: offset print, solid color (pantone), screen printing (serigraphic white), embossing

An innovative and strongly evocative image to a premium wine. The embossed text creates elegance and a sensorial impact to the hands and to the eyes reinforcing the excellence of the product.

Coevo

Design Agency: Doni & Associati

Client: Cecchi

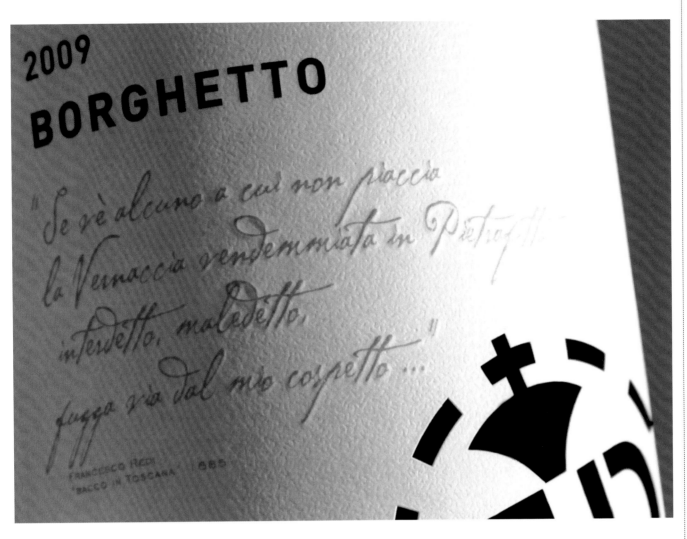

A touch of modernity to an almost classic wine label design. The punched logo dialogs with harmony with the embossed text, a creative solution with a clean and unique visual impact.

Borghetto

Design Agency: Doni & Associati

Client: Fattoria Pietrafitta

Geografia is a product series that employs geography, topography and the earth as themes. A number of different items that draw on special processing techniques and printing technology methods employed by printing companies are presented. The designers aim to attribute a shape in the form of products that can experience information, besides the printing technology disseminates information.

Metal sphere:
The latitude and longitude lines have been printed using a transparent UV ink, and appear and disappear according to the amount of light shining on them.

Leather sphere:
Each animal has been coated with UV ink. Paper with a texture reminiscent of leather has been used here.

Wood sphere:
With paper embossing and the printing of wood grain, the natural texture of trees has been recreated here.

Earth's Axis

Design Agency: Drill Design

Designer: Yusuke Hayashi, Yoko Yasunishi

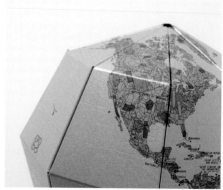

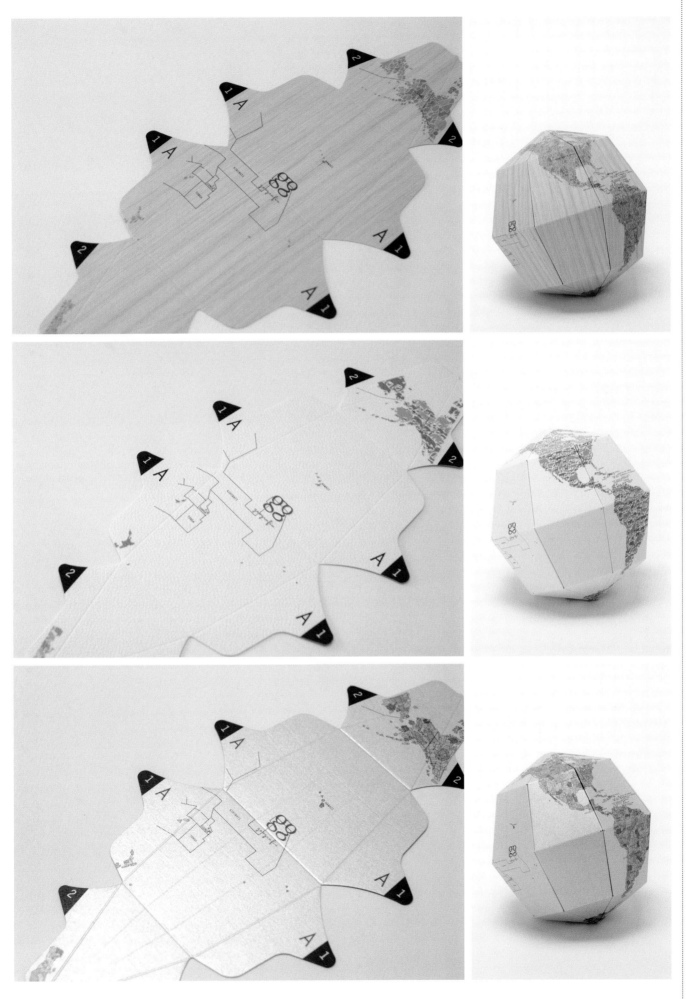

Printing technology: UV screen printing, micro flute printing, die-cutting and scoring

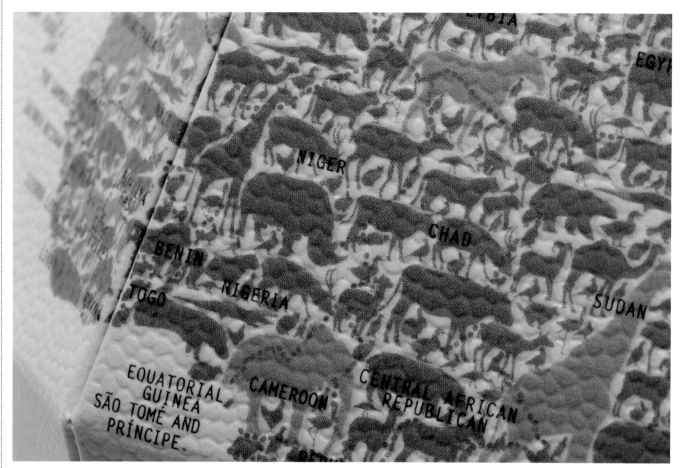
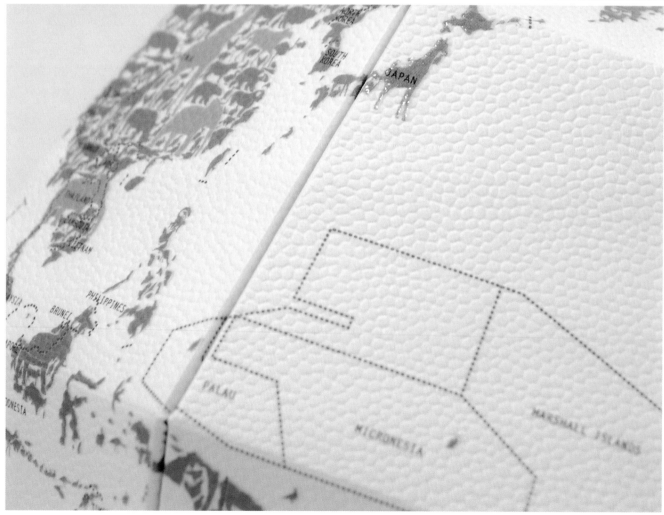

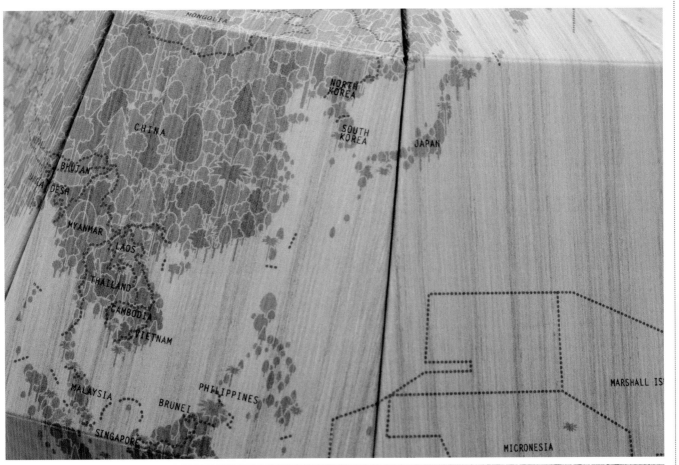

Geografia is a product series that employs geography, topography and the earth as themes. A number of different items that draw on special processing techniques and printing technology methods employed by printing companies are presented. We aim to attribute a shape in the form of products that can experience information, besides the printing technology disseminates information.

These have been created using the inline embossing technique, which allows embossing to be carried out simultaneously with printing. Screen-printing has been used, which raises the printed side up.

Planning on Paper Ring Notebook

Design Agency: Drill Design

Designer: Yusuke Hayashi, Yoko Yasunishi

The MediaBite company was formed at the time of fast expansion of social media market. The client needed designers to create the brand around their slogan "get your share of the social media cake" thus the bubble and the bite. The idea behind the MediaBite corporate identity was to form a recognizable and memorable identity while still retaining playfulness. The colors and materials were picked vibrant to communicate fresh, innovative and creative ideas which the company produced.

MediaBite

Design Agency: Kudos Studio LCC

Designer: Kresimir Kraljevic

Two inks letterpress printing on Crane's Lettra Pearl White. They use different thicks of paper — 220C or 600 gsm on principal piece (information) and 110C or 300 gsm on decorative wrapper piece and little aditional information cards.

Alejandro & Esperanza Wedding Invitation

Design Agency: El Calotipo Printing Studio

Client: Alejandro & Esperanza

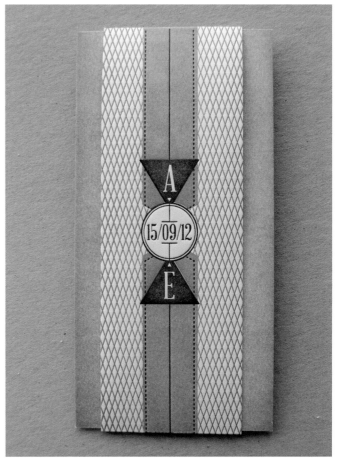

Wedding invitation 10 cm x 28 cm size on digital printing over 350 gsm paper. Custom-made envelopes on recycled paper. The join is completed with an handcrafted stamp to close the envelope looks like sealing wax.

Esther & Kiko Wedding Invitation

Design Agency: El Calotipo Printing Studio

Client: Esther & Kiko

This suite of business card and postcard was printed with a single rainbow foil on tactile book board, resulting in a number of striking color variations. The final product reflects Elana Schlenker's love of color and attention to detail and form.

Holiday Postcard Business Card

Designer: Elana Schlenker

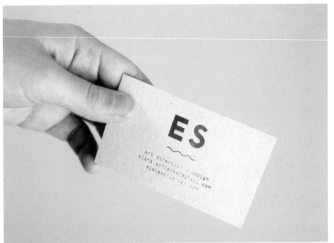

Printing technology: foil-stamping

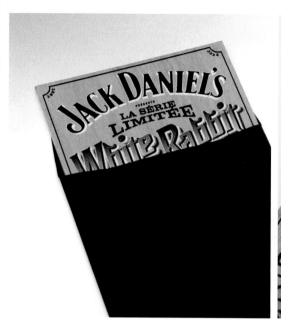

For the launch of the French special edition Jack Daniel's bottle — the White Rabbit Saloon, journalists received a wooden press kit. The rough material makes a direct link to the barrels, the spirit of Jack Daniel's.

The challenge was to print in a short amount on time detailed graphics and typography on pieces of wood, made possible with the last digital printing techniques. The result makes each press kit a one-of-a-kind traditional work of art.

White Rabbit Saloon Press Kit

Design Agency: ReflexParis

Designer: Florent Carlier

Client: Jack Daniel's

This book published as Collection of Commentaries of Prize. It is hosted by Suntory Holdings Limited, a Japanese brewing and distilling company group. The cover paper has a soft texture and an uneven surface. The title and logotype printed on the cover use a metallic hologram-foil on this paper, which changes its colors with iridescent when seen from different angles.

Suntory Prize for Social Sciences and Humanities

Design Agency: DNP Media Creations

Designer: Yoshimaru Takahashi

Client: Suntory Foundation

Award-winning book for Farstad Shipping ASA. Farstad Shipping is a major international supplier of large, modern offshore support vessels. The company's head quarter is located in Aalesund on the north west coast of Norway. In addition Farstad Shipping has offices in Aberdeen, Melbourne, Singapore and Macaé. Through a joint venture they also haS presence in Angola. The total number of shore personnel is 160 and the number of sailors is approximately 1,700. Farstad Shipping has a fleet of approximately 60 vessels and currently more vessels under construction.

The book project was developed for the companies 50th anniversary. Embossed around the cover, all fleet names throughout the companies first 50 years. The book received gold in the Norwegian advertising competition "Sterk Reklame" (Strong Advertising) in 2010.

Farstad Shipping — 50th Anniversary Book

Design Agency: Havnevik Advertising Agency

Designer: Tom Emil Olsen and Robert Dalen

Client: Farstad Shipping

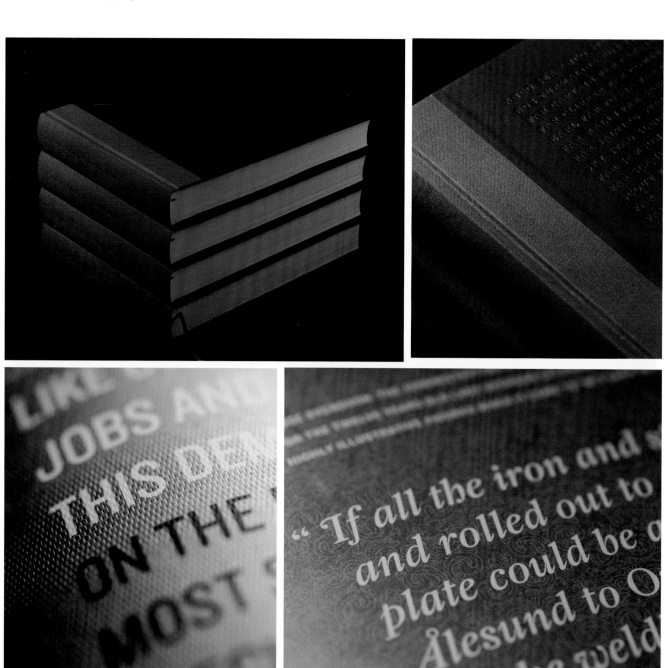

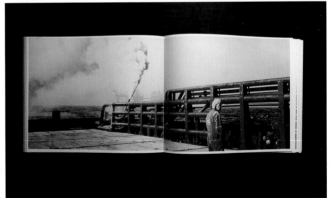
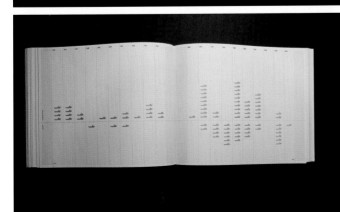
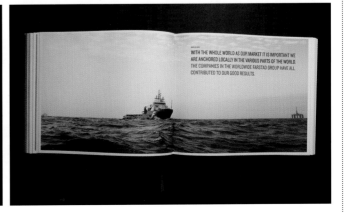

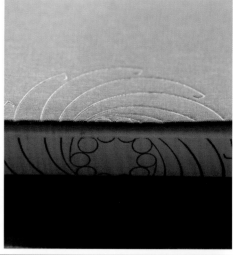

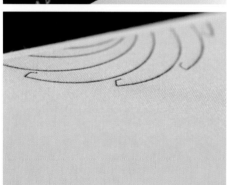

Award-winning book for Norway's leading manufacturer of gutter systems.

The Book was awarded gold in the highest ranked book-design award in Norway, "Årets vakreste bøker" (The most beautiful books of the year). The award is delivered by "Grafill" (The Norwegian organization for visual communication).

Grøvik Verk is Norwegian leading manufacturer of complete gutter systems, and the only producer of gutters in aluminum.

Grøvik gutter system is easy to install, has excellent strength, durability and is very competitive in price. The first gutter was installed in 1956 and is just as fine today.

This book project was developed for the companies 50th anniversary.

Grøvik Verk — 50th Anniversary Book

Design Agency: Havnevik Advertising Agency

Designer: Tom Emil Olsen

Client: Grøvik Verk

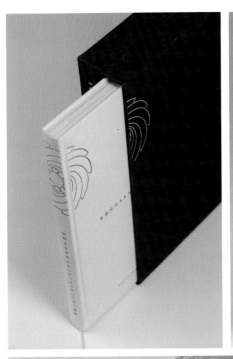

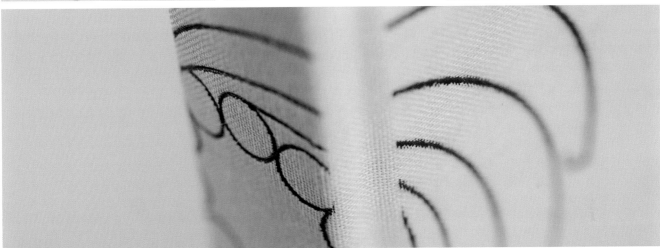
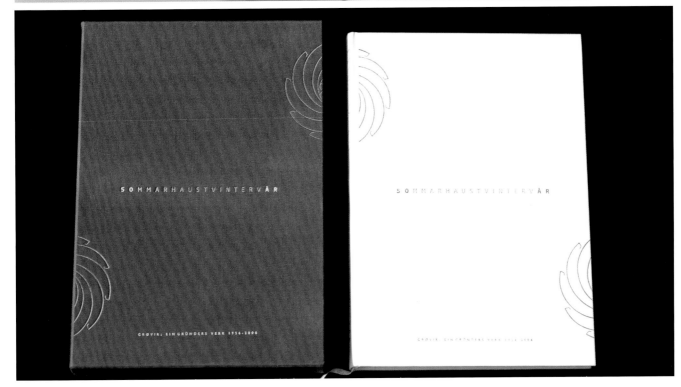

Printing technology: embossing, offset printing, laser engraving, foil stamping

JACU COFFEE ROASTERY
Established 2011

JACU COFFEE ROASTERY
Established 2011

The Jacu bird lives in South America and is known for something quite extraordinary. It flies from coffee plantation to coffee plantation and picks and eats the tastiest coffee cherries. The fruit makes its way through the bird's digestive system and the seeds of the fruit — coffee beans — come out perfectly processed. These coffee beans are among the most exclusive in the world. This story has inspired the designers, and the bird has lent its name to the new micro-roastery in Ålesund, Norway.

Jacu Coffee Roastery was established in 2011. Like the Jacu birds they pick and roast only the best beans. They look for great plantations, optimal processing, and the roasting profiles which will make the most out of each bean. They work with passion, patience, and without compromise.

Jacu Coffee Roastery, Corporate Identity

Design Agency: Havnevik Advertising Agency
Designer: Tom Emil Olsen
Client: Jacu Coffee Roastery

Printing technology: embossing, offset printing, laser engraving, foil stamping

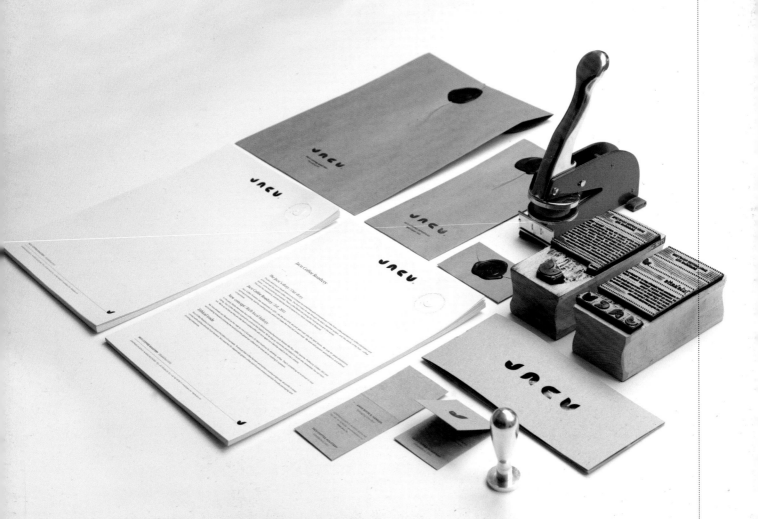

Printing technology: embossing, offset printing, laser engraving, foil stamping

Printing technology: embossing, offset printing, laser engraving, foil stamping

Printing technology: embossing, offset printing, laser engraving, foil stamping

Printing technology: embossing, offset printing, laser engraving, foil stamping

It is the limited edition vinyl cover and custom typeface for soundtrack to a catastrophic world. Features an eclectic mix of soundscapes, low frequency transmissions, sonifications, and noise patterns as catastrophe is taken as a point of departure to reveal the importance of sound and auditory perception. When the vinyl sleeve is extracted from the case, the movement of the typography in the title interacts in such a way that it creates the illusion of continuous flickering in a frequency-like way.

Catastrophic World

Designer: John Barton

Client: JumpShipRat

Printing technology: screen printing, laser cutting and foil stamping.

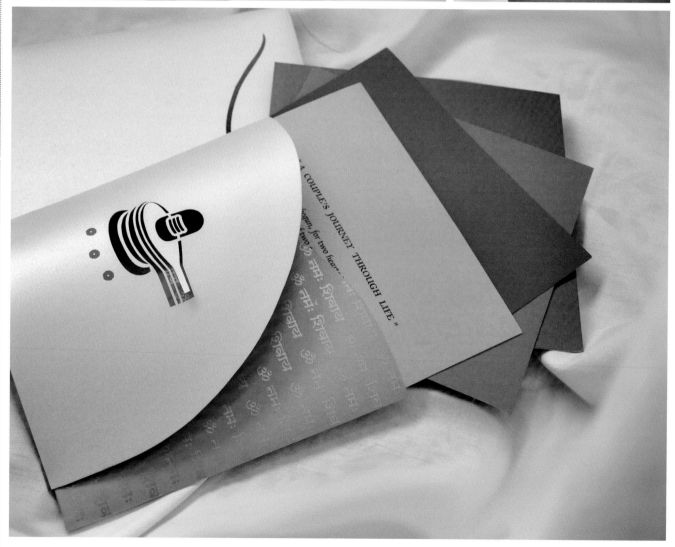

Tiny, delicate, big, bold, classy or elegant, each card has an identity of its own, symbolizing Indianess. Kaushik & Rena's philosophy has been to create custom designed cards using their knowledge of printing techniques and handcrafted elements, made in-house. To provide their clients choices that had not been available before. They enjoy a challenge when clients come up with their own ideas, which they both then, interpret creatively. Their typical client would be people who like their individual style, status, personality or interests to be reflected in their specially created stationeries. What makes their work even more special is the fact that each card is crafted by handicapped office staff whom they teach screen-printing.

Special Cards for Special Occasions.

Designer: Kaushik and Rena Shah

Client: Wedding Card Industry

The Croatian Post is one of the oldest and biggest companies in the Republic of Croatia and has recently gone through major business transformation. As a part of their transformation they needed the progress report that would instantly communicate this change. Therefore the report had to be modern, minimalistic and visually appealing to the people, business partners and potential investors — the very opposite from all their previous products. The designers used a radical approach (from their previous standards) to built this report emphasizing visual elements, large editorial and stock images and infographics to communicate their success rather than combining raw numbers and statistics tables with "business oriented" text.

Croatian Post Progress Report

Design Agency: Kudos Studio LCC

Designer: Kresimir Kraljevic

Client: Croatian Post

Due to construction work of the Graz Opera had to move their rehearsal and office rooms from a building next to the opera house to one further away. By making a virtue out of the necessity moodley devoted the program booklet to the topic of move and construction. Moodley composed the world of opera between moving boxes, wool insulator and power cables in pictures that show the aesthetically beautiful structures of these industrially produced materials. Because art is everywhere, even at construction sites, you just have to look closely. Whenever the Graz Opera can move back into their familiar environment everything will be in place: "In the end everything will be fine, and if it is not good, then it is not the end," Oscar Wilde already knew.

Graz Opera

Design Agency: moodley brand identity
Art Director: Wolfgang Niederl
Creative Director: Mike Fuisz
Designer: Wolfgang Niederl
Client: Opernhaus Graz GmbH

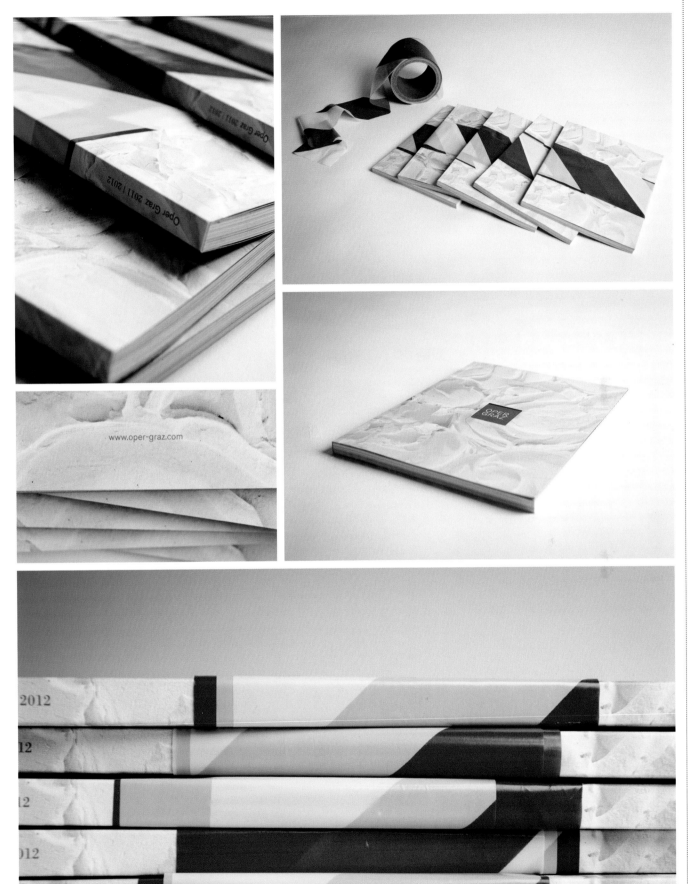

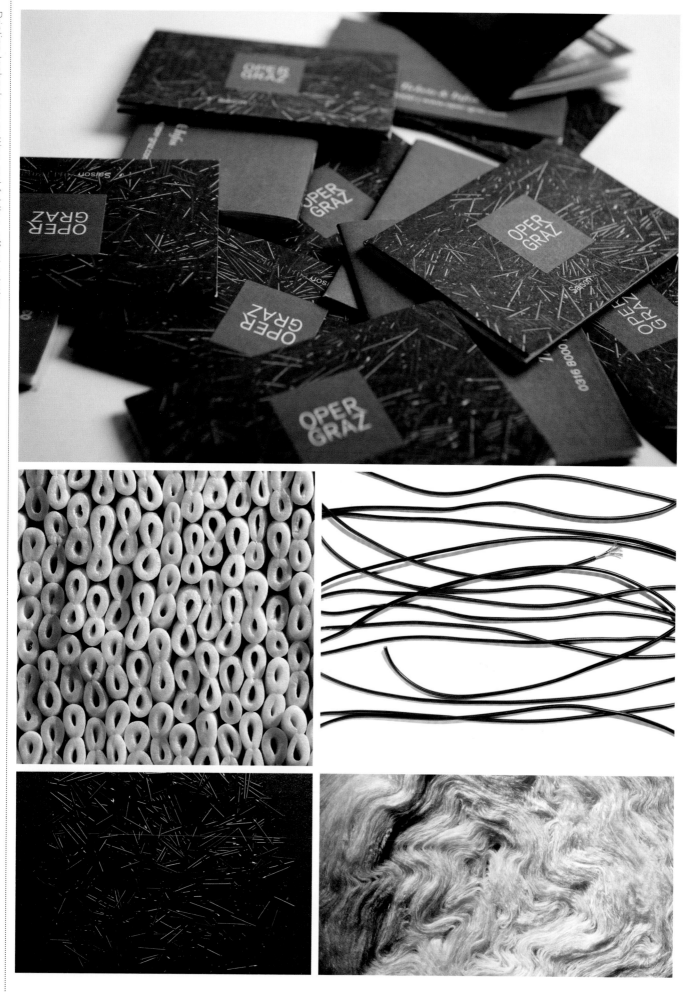

Printing technology: UV-ink, offset printing

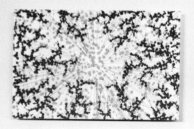

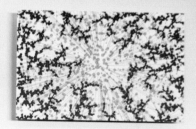

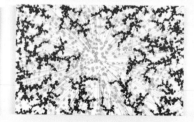

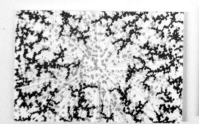

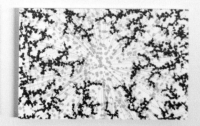

New Frontier Group is a network of globally thinking and locally acting IT companies — across state borders and very flexible! It specializes in IT solutions for Central and Eastern Europe as well as Russia. New Frontier Group combines wide and deep competence with the ability to adjust solutions to very specific markets. In addition to the logotype the generative pattern in the corporate identity — created by moodley, depicts the dynamic, multinational networking as well as the internal processes of the group.

New Frontier Group

Design Agency: moodley brand identity

Art Director: Nora Obergeschwandner

Creative Director: Gerd Schicketanz

Designer: Nora Obergeschwandner

Client: New Frontier Group

The project has used some printing processes, illustrating the beautiful book.

Beautiful

Design Agency: Shenzhen Huathink Design Co., Ltd.

Designer: Liu Yongqing

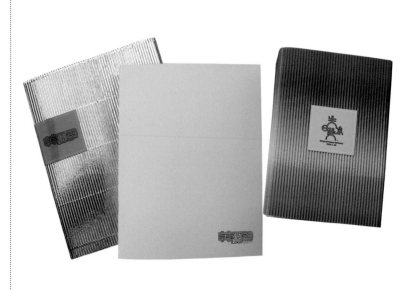

Printing technology: blue silk cloth, cardboard, hot foil stamping, silver foil, and folding

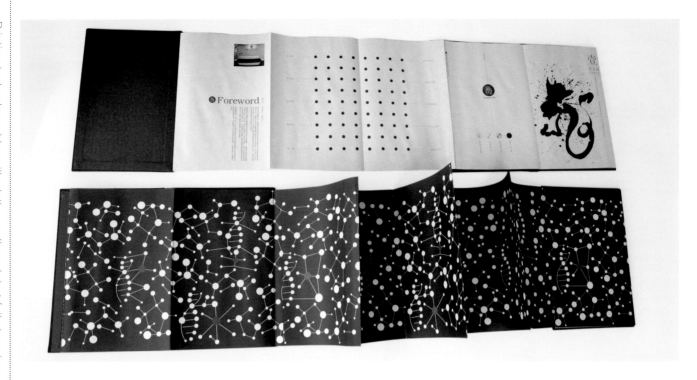

The project has used Chinese traditional folding and other printing processes. With an appeal lasting, it is elegant.

Long Mark

Design Agency: Shenzhen Huathink Design Co., Ltd.

Designer: Liu Yongqing

Client: Longmark International Holdings Limited

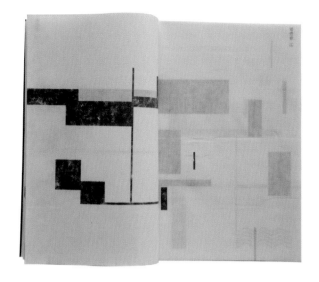

Printed on Chinese art paper, it looks charming with screen printing and other printing processes.

Memory of Jiangnan

Design Agency: Shenzhen Huathink Design Co., Ltd.
Designer: Liu Yongqing
Client: Jiangnan Yi Xiang

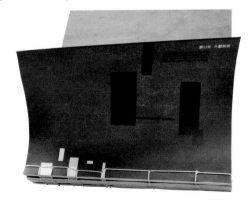

Brand design, graphics, facade and corporate materials for Betlem, an old deli grocery store at the Barcelona's Eixample district now converted into a gastro-bar specialized in quality tapas and sandwiches.

The logo is a contemporary interpretation of the modernism style of the 20th century that decorated this type of groceries in this zone of Barcelona. The ornaments are now made with the food icons and also the corporate type used is an ornamental font but less floral and more readable and clean. The materials for the stationary are chosen specifically to link to the interiorism as the bar marble tables and the metallic chairs.

Betlem Gastro Bar

Design Agency: Toormix
Designer: Ferran Mitjans, Oriol Armengou
Client: Miscel·lània Betlem

Printing technology: foil stamping, offset printing and special marble paper for the cards and letterheads menus

Printing technology: foil stamping, offset printing and special marble paper for the cards and letterheads menus

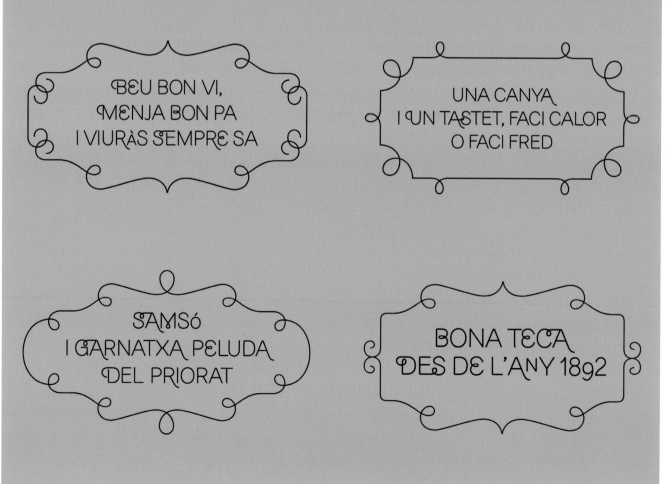

Printing technology: foil stamping, offset printing and special marble paper for the cards and letterheads menus

It is a rebranding and new corporate material design, creation of its new graphic code and communication materials for our favorite printers in Barcelona on the occasion of its tenth anniversary. The graphic exercise plays with a "K" the letter then takes more protagonism in the naming. The shape is a parallelism of a wood type used in old printers. The graphic applications always have a "K" pattern in different printed ways.

Grafiko

Design Agency: Toormix

Designer: Ferran Mitjans, Oriol Armengou

Client: Grafiko

It is a book on artist Josée Dubeau. The cover features one of her work and uses debossing and a solid metallic color.

Contour

Designer: Simon Guibord
Client: AXENÉO7

Printing technology: white offset ink and overlay of paper

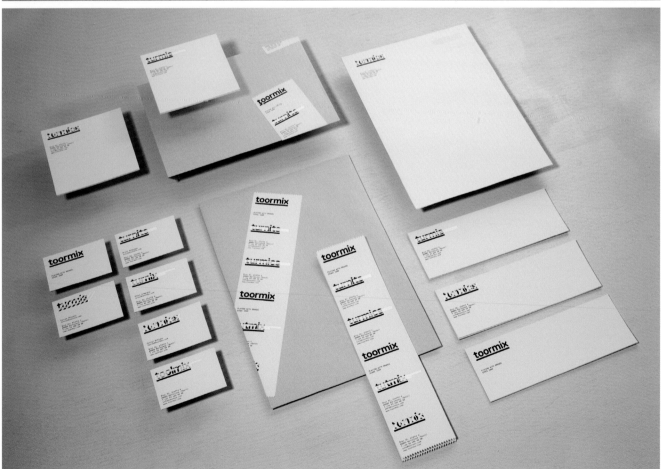

New identity to celebrate the first 10 years of the studio. It is a new brand and a new corporate color, the yellow. The designers wanted a bigger visual impact so they choose the yellow and black combination for all the pieces. Also in the print and use of paper they played with superposition of these both colors, for example for the cards and the booknotes. This notebook is the designers' daily work tool and is for that reason that they wanted an object to give to their clients and contacts. Also they titled this notebook as "brainstoormix".

Also, for their tenth anniversary they did a collection of posters based on the new image and playing with ideas over the 10 first years.

And finally the designers also renewed their website based on this new identity and colors.

Toormix Stationery

Design Agency: Toormix

Designer: Ferran Mitjans, Oriol Armengou

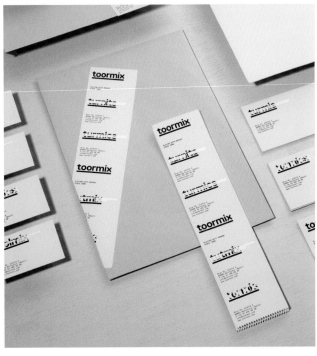

Printing technology: foil stamping

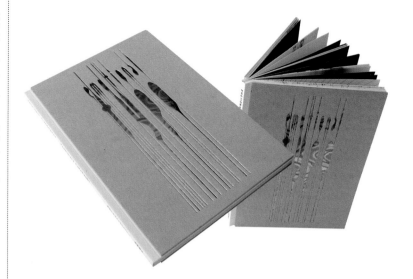
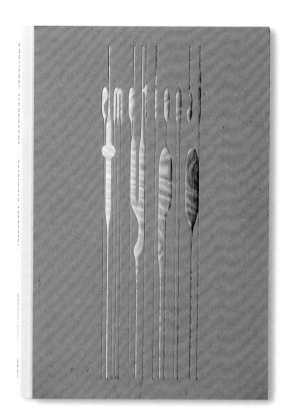
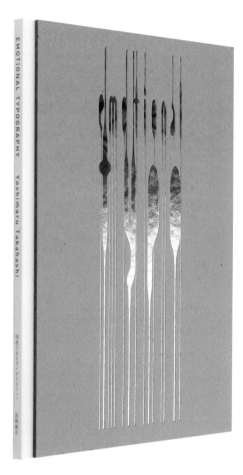

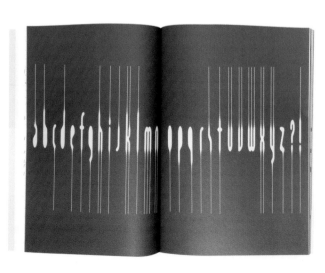

It is typography design book.

The cover is a kind of paper that has a thickness of 3mm.

The title and logotype printed on the cover use a rainbow-foil on this paper.

Emotional Typography

Designer: Yoshimaru Takahashi

Client: ddd gallery

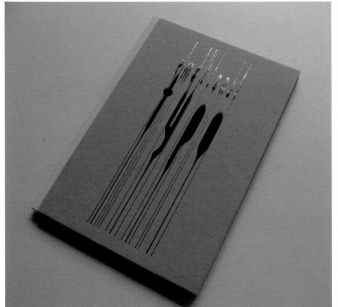

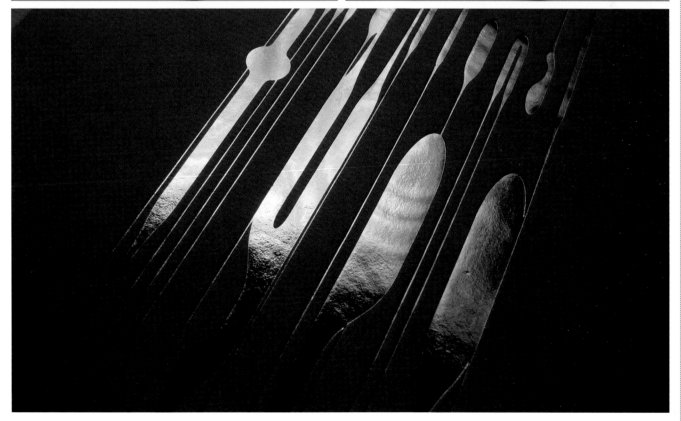

QT Hotels, Gold Coast — Fixx Cafe & Bazaar Restaurant part of the designers branding work for QT Hotel, Gold Coast, included creating identity and print materials for its new cafe and restaurant. Complementing the interiors nostalgic 50s/60s Beach Glam theme, raw materials such as hessian and timber, further enhanced with screen printing and laser etching were used to give a distinct, tactile and highly memorable feel, appropriate to the venues.

A Bazaar Fixx at QT

Design Agency: THERE

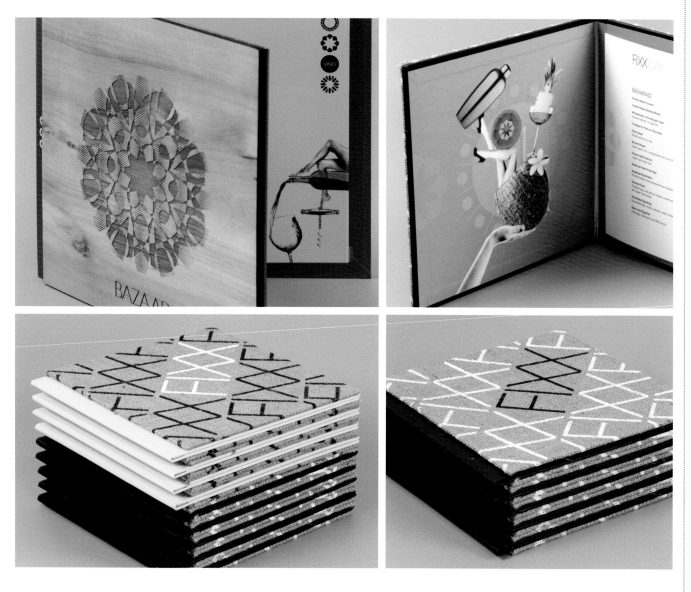

Contributors

Analogue

They are Analogue…

They use design, they use digital, they use art direction, copywriting, and good old-fashioned perspiration. They immerse themselves in culture, taking inspiration from all aspects of life, always looking forward and striving for perfection… Over the years they've collaborated with a wide range of clients, each one with their own unique personalities and individual requirements, but the results have always been the same: They help brands stand out from the ordinary. They help them become extraordinary.

Andres Reisinger

Andres Reisinger is an argentinean designer. He started his professional career at Metropolitan Design Center, where he mostly worked doing editorial, print, event and corporate spaces. Thanks to his outstanding performance in this place, and also his well-known development as student at Universidad de Buenos Aires, he was called to join the RDYA's team, where he has worked with storyboards, style-frames, corporate branding, identification systems, commercials and network branding. Now he is working at PLENTY for Fox, Axn, Nick, and more recognized brands.

Artifacture

Artifacture is a small design–build studio in the south of downtown Dallas that specializes in laser cutting. They are makers, creators, tinkerers, programmers, yeah, they're nerds. Whether it's a prototype of your pet project or a desire to experiment with new materials, if you can think it, they can make it. If you can't think it, they can still make it.

Artifacture's experimental workshop utilizes traditional and high-tech tools to create custom solutions for their clients. With 209 m^2 of workspace, their studio comes complete with all the tools, gizmos, and gadgets a creative thinker would drool over. They don't just love their trade–they live it–in their attached living area. So when inspiration moves people, they're ready to strike.

Arvi

Arvi is an avid snowboarder and wishes to one day be a snowboard bum living in a small cottage in Zermatt, Switzerland. But for the time being, Arvi can be found in San Francisco working with Hot Studio. Arvi has worked with a number of companies from small start-ups to Fortune 500 on projects ranging from corporate identity, annual reports, websites, and environmental design to corporate branding and brand experiences.

Arvi's work has been recognized by a number of design shows and publications including: Graphis, AR100, Communication Arts, Print, HOW, the BoNE show (Best of New England), Graphic Design USA, The Big Book of Logos, Best of Brochure Design, 1,000 Greetings, Fingerprint 2 and most recently Just Design, a book that celebrates and explores socially conscious design for critical causes. Arvi has judged for various national design shows and has spoken at various venues on the subject of the power of design for social good.

ARE WE DESIGNER

ARE WE DESIGNER, they are 152 innovative and exciting design projects, 56 daring and pleased clients, having 12 years of design experperience. For all projects they seek custom-fitted design solutions that are effective and flexible, always boldly searching for what's new.

Some call it an "agency", others "designstudio". ARE WE DESIGNER themselves say: "European Design Hasardeure" (Risktakers). Why? Because they don't fit into a run-of-the-mill design agency drawer. They're much smaller. They are more efficient. So rest assured: they know the ropes, and love what they do. So be bold, because they are, too!

Annalisa Kleinschmidt

Annalisa Kleinschmidt is from Santa Fe, New Mexico and has studied communication design at Rocky Mountain College of art + design in Denver, Colorado. Her work ranges from print and graphic design to photography. A major part of her design philosophy is to fill the gap between science and design through process, thought, and communication. "Science is a way of thinking much more than it is a body of knowledge" and she is a design thinker.

Become

The Stockholm based agency Become is an intelligence driven design studio with focus on adding distinct visual charisma to brands, companies and products. Providing tailored design solutions for both established brands and small start-up enterprises, Become is constantly aiming to reinvent the obvious as we regard expected solutions to be uninspiring. Although the agency's body of work spans over a wide range of disciplines, the specialist area is the one of high-end book design. Past work includes the mammoth 16 kg official autobiography about the famous footballer Pelé, a glass covered giant limited edition book of New York selling for $15,000, skis for HEAD Sports, and the praised 2012 book cover for Neil Strauss "Everyone Loves You When You're Dead" Swedish edition.

All Become's work is art directed by Henrik Persson, who started the agency back in 2007. Henrik was a 27 year old Swede at the time who was priory to forming the company Creative / Art Director at the Swedish design agency Maräng for three years, and with educational roots from London's Central St. Martins where he held a Bachelor in Graphic Design from 2004.

Become is today operating in an international marketplace, with business engagements stretching from the Austrian Alps to the deserts of Dubai.

Benoit Ollive

Benoit Ollive is a 25 years old graphic designer and artist. Graduated from the fine arts in the south of France, he quickly moved to London to study product design and started his own creative studio called Graphicfury in 2010. His work is inspired by urban trends, social networks and the underlying concepts of alternative cultures. Most of the time, Benoit chooses "DIY" (Do It Yourself) solutions and makes his own ink, binding or paper. He has been always exploring mixed media, questioning the principles of design and connecting people through narrative. Benoit is now art director at DOJO Werbeagentur Berlin and still work at Graphicfury. He recently had a solo show in Hong Kong at the Space Gallery in June 2012 and in Brighton at the Royal Pavillon Museum in October 2013.

Bond

Bond is a Helsinki based creative agency specialized in branding and design. Bond's approach is characterized by attention to detail and a passion for quality.

Bold-inc

Bold-inc are a passionate new breed of thinkers within the branding and packaging industry, offering down-to-earth, honest, cut through thinking, giving brands a life and a future.

They understand the challenges and obstacles brands face. Every brand and project is unique and therefore don't apply a formulaic approach. They challenge, make bold moves and disrupt the category (if required) to get the best result, delivering truly memorable design.

With a combination of creative expertise, lateral thinking and a pinch of good old fashioned common sense, Bold-inc prides itself on the fact that they "think before they inc".

Bunch

Bunch is a leading creative design studio offering a diverse range of work, including identity, literature, editorial, digital and motion. Established in 2002 with an international reach, from London to Zagreb Bunch has an in-house team of specialists to deliver intelligent and innovative cross-platform solutions of communication design. Over the years they have been commissioned by many blue chip companies as well as younger brands and artistic industries. Building an impressive client base that covers many styles and disciplines, such as BBC, Nike, Diesel, Sony, Sky, Red Bull and etc.

Bureau Rabensteiner

Bureau Rabensteiner is an Austrian design studio specialized in creative direction and graphic design. They combine strategic thinking with branding and photography therefore they are able to transport more than just design but a whole company spirit on different channels. The Bureau wants to give an insight in the company life and surroundings. They share the things that inspire them, which emerged to be a good way to stay in permanent contact with other designers and interested people all around the world. Not at least it gives them the opportunity to connect with the right kind of clients, who match their style and thinking and bring the most interesting new projects and challenges.

Cliche Letterpress Studio

Almost disappeared some time ago the high printing is in great demand again as one of the methods for multiplying texts and pictures. Their workshop is one among those who renew the printing techniques in its new quality. And if in America this kind of printing is widely spread and popular in Russia, their studio becomes a pioneer in this field.

In their opinion, only the high printing allows to show the fullness of an artistic image that one can print on paper. Raised stamp, led with sculptural beauty, wonderful combination of structures and pictures, real sense of touch ability and naturalism — all this is the high press. And that is what the modern set and digital printing are missing.

Add a high-quality 100% cotton paper, which has inner warmth and charm, and you get really "live and sensitive" stamp.

But you must not forget people, without whom the stamp would stay no more than a stamp of an indifferent machine. Loving so much their craft they create that wonderful mystery of hand-printing, putting a parcel of their souls in every stamp.

Cue

Cue is a brand design company that helps clients clarify and build meaning into their brands. Navigating from strategy to creative expression requires grounded thinking, intuition and understanding. But more than that, it requires taking ownership of a meaningful idea in the name of a brand. Cue works collaboratively with clients to understand the underlying business challenges and uncover the meaning and value that defines them. The idea may come to life through brand identity, packaging, collateral, print, web or an environment. Cue uses design as a tool to unleash a brand's true identity.

Doni & Associati

Doni & Associati is one of the very few agencies worldwide focused on wine design and communication. For over 20 years it has been working for the most prestigious and established companies, as well as for some of the most innovative young wineries worldwide.

It has been always searching for the emotional impact of a wine brand image that can have a significant impact on customer choice and a premium quality packaging design that can represent a profitable investment for the producer.

In each project aiming at maximizing strategic discipline, creative excellence, production values and efficiency, Doni & Associati develops corporate identity, packaging design and marketing materials.

DMWORKROOM

DMWORKROOM is a London based design studio set up by Denis Mallet, a French graphic designer and art director with a contemporary and multidisciplinary vision. They believe graphic design is a dialogue between the client, designer and the audience. They are interested in brand design and they believe that the most successful work can only be for brands that also share their approach. Brands that are driven to be about individual preference, to be entrepreneurial in their attitude and that esteem the crafted qualities within the products they offer. To these they bring creativity, inventiveness, fresh design ideas, energy, dedication and sensitivity.

Drill Design

Design studio of Yusuke Hayashi and Yoko Yasunishi.

After their college education, both studied design and founded DRILL DESIGN in 2000. They give total project directions in various fields, such as product design, graphic design, and interior design. The products which are mainly everyday objects such as furniture, stationeries, gardening tools, and kitchen utensils, are available for sale at countries around the world, including the MoMA Store in New York. They have won numerous awards, such as the prestigious Red Dot Design Award (in Germany) and the Good Design Special Award.

Elana Schlenker

Elana Schlenker is a graphic designer specializing in print, identity and interactive work. Drawing from a background in book and magazine design, Elana brings an editorial eye to the organization and implementation of each project. Her work is driven by an idea-led approach, which marries conceptual thinking with formal typographic principles and a contemporary sensibility. Formerly an art director at Conde Nast, Elana is currently part of the design department at Princeton Architectural Press. She also works independently with clients across a wide range of industries, and publishes and designs gratuitous type — an occasional magazine of graphic design and typography.

El Calotipo Printing Studio

El Calotipo Printing Studio is a craft workshop that fuses traditional graphic processes with new technologies in design. They investigate new materials with special finishes to provide publishing products and original artworks.They produce and comercialize selected editorial products, also teaching Book Arts workshops (bookbinding, engraving, silkscreen and letterpress) and offers a rental service facilities for artists, designers and creatives.

End of Work

End of Work is an Australian-based design consultancy, which prides itself on creating value enhancement for its clients through creative solutions anchored by a strategic approach. They possess a passion to distinguish their clients' brands in the marketplace, through innovative ideas and beautifully crafted design.

They have produced a wide range of acclaimed work for a broad client base — from global mining companies to boutique restaurants. While End of Work's portfolio is certainly diverse, every piece represents the same combination of strategic thinking and creativity. It's this trademark obsession with smart ideas and meaningful delivery that sets End of Work apart.

Eva Slade

Eva Slade is a design studio based in London specializing in letterpress stationery. At the heart of each project is passion for paper, print and pattern.

Fluid

Fluid was launched in 1995 and first established its design and branding credentials within the music industry. As the company has grown the entertainment and youth culture theme has continued and to date Fluid enjoy a global reputation as a leading design agency, offering a fully comprehensive creative service. Fluid is not only their name, but also their branding philosophy. They have no house style, and their studio represents a rich pool of outstanding talent from a wide range of artistic disciplines. At every stage of the creative process they guarantee flexible, comprehensive design solutions that bear testament to the breadth of their experience and versatility of style. The company ethos "Never Not Creating" ensures every project they undertake benefits from an ingrained passion and cultural knowledge, earning Fluid a renowned reputation for consistently evolving creativity. From initial brainstorming through to development, solution and implementation, Fluid employs a strategic approach to the design process.

Firebelly Design

Firebelly is committed to cultivating connections between human beings and ideas, inspiring conscious thought and action. As early advocates for socially responsible design, they pioneered an ethic that values honesty, empathy and Good Design for Good Reason™.

filthymedia

filthymedia was born in 2004 out of a passion for design. They turn this passion for design into creative and commercial success for their clients.

From brand creation to concepts and identity, they pride ourselves on delivering effective design solutions and their tailored approach to every brief allows them to produce truly memorable and engaging work.

They work across a wide range of clients and sectors, from music to fashion creating big campaigns, small projects and everything in between. Their portfolio includes graphic design, art direction, branding, typography, web & motion design, illustration, photography and copywriting.

Florent Carlier

Florent Carlier is a graphic designer and art director currently living in Paris. He has worked in agencies in Paris and Tokyo, and also as a freelancer for independent musical groups, and cultural events in France and Belgium. He finds his inspiration into a variety of creative field, such as typography, photography, music, fashion…

In his print design projects, the paper or materials and techniques he uses are part of the concept: the message and the way to say it are equally relevant.

Ghost

Ghost is a multidisciplinary design studio with emphasis on visual identities, web development and product design, driven by passionate craftsmen aiming to create vibrant, sustainable and functional solutions.

Ghost is founded and run by designers. They work for clients who value creative, practical ideas and appreciate working directly with the designers. Ghost aims to have a continuous interaction (involvement and communication) between client and designer from project start to delivery.

HLD / hyperlocaldesign

Hyperlocaldesign is a creation positioning which is founded on design — as its key tool in the exclusive brand building. A wise and multi-cellular design proposal, holding both the physical and digital world. A process from essence to the final result.

A new way to look at the creation job, valuing the techno-biological symbiosis, through a flexible and unlimited way.

Havnevik Advertising Agency

Havnevik Advertising Agency is one of the leading advertising and design agencies in north-western Norway. The agency's varied client portfolio includes some of the country's leading companies. Havnevik is an award-winning agency that provides services in a number of fields, including public relations and communication, advertising, identity design/branding, graphic design, web design etc.

hat-trick design

Established in 2001, hat-trick is a multi-disciplinary design company working for a wide variety of clients. Their aim is to provide their clients with the highest standard of creative design and project management. They believe that the best way to achieve this is for the directors to be very hands-on and to lead the jobs from the front. Their passion is to produce work that achieves its targets by creating memorable, engaging ideas that are noticed and effective. They have a very broad experience in a wide variety of sectors. In other words they bring to all their projects fresh and objective creative thinking, not formulaic answers. They are currently ranked No.1 in the design week creative survey.

Their approach to branding is creative, flexible and involving. They're not afraid to ask difficult questions or to give honest answers. They like to feel their clients are part of the team. Their experience shows them that successful projects depend on commitment, clear communication and a shared understanding of the processes and issues involved.

Happycentro

The studio began in 1998 in Verona, the romantic city of Romeo and Juliet. In recent years, they have worked with both big clients and small startups, for local agencies and major international companies. Their approach to design is always the same: designing a logo, an advertising page, a wall or directing a commercial offers the same opportunity to deal with a problem to solve. Time by time they have become quite good at it. Mixing complexity, order and fatigue is their formula for beauty. Always, in addition to the commissioned work, they spend plenty of energy in research and testing. They work on visual art, typography, graphic design, illustration, animation, film direction and music. They like contamination between creative disciplines and diversity in general. They don't like doing the same thing twice and prefer to go beyond what they are already able to do. It is tiring but satisfying.

Hear Agency

Hear Agency is mid-western U.S. based marketing and design firm specializing in digital deployment and getting brands and businesses on the digital map. Hear Agency likes to approach design problems head-on and turn them into gorgeous solutions. Hear Agency helps you leverage the power and connectivity of the internet to reach potential lifelong customers the world over.

Holger Lippmann

Holger Lippmann comes pretty straight from visual art. He started painting as teenager, did evening courses, studied at the art academy Dresden and did his diploma for sculpture. Then he was a 2-years master student, studied as a postgraduate in Stuttgart, Paris and later moved to Brooklyn, New York, where he attended a first computer graphics internship at the Institute of Technology.

When he came back to Germany in 1994, he moved to Berlin, right in the middle of the electronic music boom, he started pretty radically to work with computers (only), he knew he had to work the same way in which the music he was listening to was made .

He passed another 1 year multimedia design education. Since then, none of the fascination of working with software and internet based networks has abated. He never felt any reason to move back to paints and traditional artistic work.

Anyway, deep inside he was a painter and he always is. So he sees his work best described in the traditional context of painting. The focus is on the development of an image and color composition.

Inksurge

Inksurge is the blend of Rex Advincula and Joyce Tai. Both graduates of computer courses, (Rex is a B.S. in Information Technology, Joyce is B.S. Computer Science and B.A. Computer Arts in San Francisco), they met in their day jobs for a local IT firm — where they both got bored in their day to day work and decided to take the leap: creating a backyard design studio, meeting in coffee shops, and working from computers stationed in their parents' homes — forming Inksurge.

Driven by a surge of passion for design, experimentation, inking up new ideas and concocting unique imagery, the tandem has become a household icon in the local design sphere, also making waves globally — with work featured in several design books and publications in Singapore, United States, Spain, and Germany. Always wide awake, Inksurge is sure to be brewing up something new.

Ineo Designlab®

Ineo Designlab® is a multidisciplinary design agency working with corporate and brand identities as well as graphic design for a broad range of national and international clients.

Mixing creative competences with strategic thinking, Ineo Designlab helps businesses bridge the gap between visual ideas and reality.

Jacob Jensen Design

Jacob Jensen Design is Scandinavia's oldest and most award winning design consultancy, established in Denmark in 1958 by Jacob Jensen. With more than 50 years of insight into innovation, psychology, sustainability and technology, the studio provides strategic design solutions for clients all over the world.

Graphic designer Nigel Hopwood became head of the studio's graphic department in 1990. He is responsible for a great number of the studio's iconic graphic works, including development of corporate identities, logotype s, design manuals, packaging and graphic user interface among others. Nigel Hopwood's work shows the essence of the quiet Scandinavian beauty.

John Barton

Liverpool-born John Barton is a recent graduate who now freelances at a number of established studios in the UK and regularly collaborates with other practitioners on independent projects.

Joseph Rossi

Joseph Rossi has established the design studio, which focuses on corporate identity design, packaging design and advertising etc.

Kalimera

Kalimera is a multidisciplinary studio that develops efficient communication systems and complex graphic identities based on a strategic and efficient investigation and research. Established in Reggio Emilia on 1996 their main aim is to manage each project with the best of their creative potential. Main projects: Vodafone, Marazzi, Sky, Credem, Telecom Italia, Barilla, Kellogg's, Gruppo Max Mara, Absolut Vodka, Coca Cola, Biennale Architettura Venezia.

KucHen

KucHen is the advertising production company established by art director Tatsuo Ebina in 1986. After being established, they have dealt with various advertising, graphic, web, film, product, branding and spatial projects.

Karolin Schnoor

Karolin Schnoor is a German designer and illustrator who lives and works in London where she completed her degree at the London College of Communications. Her love of screenprinting and old folk illustrations informs her practice. She favours bold colour palettes, silhouettes and the use of pattern to create layers and structure.

Kaushik and Rena Shah

Kaushik and Rena Shah have run a small screen printing press in Mumbai, India for over 30 years of their married life, where they create beautiful, handmade wedding cards. Their clientele is scattered across the globe in UK, USA, Canada and India. Indian weddings are about pomp, show and grandeur. Each client has a unique demand, which they both have matched creatively. With Kaushik, an art-school dropout and Rena being a textile designer their cards epitomize culture, bright colours, texture, ethnic patterns, modern designs and usability. They combine screen-printing with a touch of innovation. An eccentric clientele has made bizarre demands of cards that they have successfully created using wooden shavings or a dash of tealeaves. Keeping up with the classy and simpler tastes of the newer generations they have a collection of designs that makes it impossible to choose from. Currently, they are developing a new range of cards inspired from unconventional materials.

Kudos Studio

Kudos Studio is a visual communications agency with a primary focus on providing graphics design and IT services. The agency was established in 2009. Due to dynamic nature, size and agility of the agency, the Kudos' team is always in touch with changing needs, business values and established rules of business communication of clients toward media and other audiences. Kudos' team is comprised of 6 experts and local network of professionals in differing visual communication areas. Their creativity and experience to a large extent add to success of the projects and overall business efficiency of the agency. Every expert and professional working in Kudos has a specialist training and knowledge which they constantly upgrade and expand. The Kudos Studio LCC provides their clients with innovative solutions and content by creating and executing crative and ussable communication initiatives.

Kylie McIntyre

Kylie McIntyre also known as Peachy Flamingo is a Sydney based graphic designer. Not only is Kylie McIntyre an art director / graphic designer, but also she is an all round creative dabbling in craft, photography and illustration — all to which plays apart in her aesthetic. Over her short career span Kylie McIntyre has worked on a variety of clients over a spectrum of industries, in areas specific to print & digital, branding, advertising, social media & event design.

Outside of work Kylie McIntyre also runs her own freelance business "Peachy Flamingo" in which she has taken on pro bono work for charities and contributed to art and music festivals such as the A4 Paper Festival, all to which helps her develop as a creative. Kylie McIntyre's willingness to experiment and push the boundaries in her work, means that things can only get "peachy" from here on in.

La Tigre

La Tigre is an independent media studio in Milan, directed by the designers Luisa Milani and Walter Molteni. Since its opening in 2009, La Tigre takes on a wide variety of projects of different nature, such as web, printing, branding, editing and illustrating, always exploring original and alternative solutions.

LSDK

LSDK Design + Konzeption was founded in 2009 as a multi-disciplinary creative company with office in Suttgart. Their fields are branding, graphic, web design & consulting.

Mama's Sauce

Mama's Sauce was founded by 3 Italians and named by a happy accident while mixing a chunky red ink in a cobalt blue bowl. While they may not mix their inks in a kitchen anymore, they give their spot color print work the same attention that an Italian Mama gives her Sunday sauce.

Marnich Associates

Marnich Associates is a design and communication consultancy based in Barcelona. They believe in simplicity and clarity. Their clients range from small restaurants, independent publishers and music festivals to large corporations, banks and museums. They hope their work is of interest to you.

Mind Design

Mind Design is an independent London based graphic design studio founded in 1999 by RCA graduate Holger Jacobs. Mind Design focuses on integrated design which combines corporate identity, print, web and interior design. Mind Design works for a wide range of clients across various sectors, from start-ups to established companies.

Their philosophy and approach is based on a passion for craftsmanship and typography. They offer practical and friendly design solutions and believe that content and form are inseparable. Every project is seen as a new challenge and they never follow an already established graphic house style.

moodley brand identity

moodley brand identity is an owner-led, award-winning strategic design agency with offices in Vienna, Graz and Shanghai. Since 1999 they have worked together with their customers to develop corporate and product brands which live, breathe and grow. They believe their key contribution is to analyze complex requirements and develop simple, smart solutions with emotional appeal whether corporate start-up, product launch or brand positioning. Their team currently consists of more than 40 employees from 7 different countries.

Murmure

Murmure is a french creative agency founded in 2010. Murmure focuses across art direction, branding, identity print & digital. Proudly led by julien alirol & paul ressencourt.

MyORB

MyORB is a design studio based in New York. They create identities, websites, typefaces, books and brochures, design exhibitions, produce short films, and make other things. The aim of their work is to be curious, intriguing and playful. The goal of their practice is to create mutually beneficial relationships with people and organizations whose work they admire. Clients range from a variety of industries such as art & culture, architecture, media and fashion.

Naoki Ikegami

Naoki Ikegami was born in Kobe, Japan in 1978. He graduated from Nihon University College of Science and Technology in 2002. After graduation, he worked for a business promotion company as a designer for 2 years, for a package design company for 2 years. He established KOTOHOGI DESIGN, a design office in 2006. Lives and works in Tokyo, Japan. He is a member of Japan Graphic Designers Association.

Nevertheless

Nevertheless is a multidisciplinary design studio founded in 2011 by Dimitris Deligiannis and Stavros Kypraios, working in the fields of art direction, graphic and digital design, illustration, installation, multimedia and animation. Small as they are, they are flexible. They're top priority is to provide excellent and creative visual communication in each and every project they undertake. No matter what, they love to experiment with new methods and techniques, getting their hands dirty and push themselves to the limits.

OddFischlein

OddFischlein is a design agency situated in Aarhus, Denmark. OddFischlein works with visual identity creation. The primary focus areas are corporate identity, font design, illustration and web design. Their approach to design is a crisp combination of method, crafts, ideas and commitment. The goal is to have their visual solutions include the recipient. They want to give clients the opportunity to have an experience of insight and not just recognizability. Their workflow is divided into three phases: destroy, rebuild and enjoy. This way, they ensure that the design solution meets the customer's needs supplied with innovative ideas.

otto design lab.

otto design lab. is formed with the three young men based in Osaka, Japan. They design in various directions and even involved in such as company's management and operation etc. Their work follows the concept which is to do branding in making the best use of the client's potential ability.They add design to social activities to not only make benefit but also to solve any problems.

Raquel Quevedo Studio

Raquel Quevedo Studio is an open/collaborating studio founded in 2011 in Barcelona, and now based on Berlin. The studio works are focused on identity and packaging design with a very strong art direction on typography and printing results.

In the work featured the studio collaborates with Diego Ramos as a creative director. Diego Ramos is an industrial designer with a background on experimental design based on Barcelona.

Reut Ashkenazy

Reut Ashkenazy is a graphic designer based in Tel Aviv, Israel. She graduate from Shenkar, college for engineering and design in 2009, and hold B. design degree of visual communication. Reut opened the design studio in 2010, which specialize in branding and print design (visual identity, logo design, packaging, posters etc.), Art direction and production (concept development, casting, production, art and styling on set and design) and Interactive design, the studio works with a wide range of global clients, many within the fashion and cultural industry. Among the studio clients you can find Adidas AG (Germany), Adidas (Israel), Mark Betty Publishing (USA), PayPro Global (Canada) etc.

Rowan Toselli

Rowan Toselli was born in 1988 and spent most of his childhood counting acorns and drawing with wax crayons at a Waldorf school. He studied Graphic Design at the University of Johannesburg, before joining the Nicework team as a designer. He currently works as a freelance designer and art director and has worked for various agencies including Hello World, Agency and Yellowwood. He enjoys climbing, character design and creating objects out of wood. Most of the time, you can find him drinking tea and doodling monsters.

Ruchi Shah

Ruchi Shah is an illustrator currently practicing in London and Mumbai. With a flair for screen-printing she also works with different ubiquitous materials.

Previously, she has had 3 years of experience in design research, worked as a senior visual designer and as an illustrator with over 10 successfully published books. With a specialization in screen-printing, she has custom designed limited edition books for clients. After having worked with corporate, start-ups, non-profit organizations and publishing houses, her focus revolves around making 3D "installative illustrations" and her family run screen-printing business which generates custom designed cards for clients across the globe.

Ryszard Bienert

Ryszard Bienert — Graphic Designer/Art Director

Born in 1976, graduated in 2002 as master of fine arts at fine arts academy in Poznan. He has worked as a graphic designer since 1996. During his career he worked with many advertising agencies and design houses in poland both as an art director and on freelance basis. In 2006, he co-founded a graphic design consultancy — 3group. He has designed publications for artists in Galeria Piekary since 2005. In 2008, he was awarded european design award merit in Stockholm for "Waiting for the Barbarians" catalogue. In 2010, he was awarded european design award "silver award" in Rotterdam for "Goalkeeper Forever" catalogue. In 2012, he was awarded european design award "gold award" in Helsinki for "Dobosz" catalogue. Many of his works reflect his strong typographical fascination.

Sagmeister & Walsh

19 years after the founding of Sagmeister Inc., they are renaming the company into Sagmeister & Walsh. View the announcement card here. Stefan Sagmeister was born in Austria and lives and works in New York. He has worked for the Rolling Stones, The Talking Heads, Lou Reed, The Guggenheim Museum and Levis. Exhibitions on Sagmeister's work have been mounted in New York, Philadelphia, Tokyo, Osaka, Seoul, Paris, Lausanne, Zurich, Vienna, Prague, Cologne and Berlin.

Jessica Walsh is a multidisciplinary designer working in New York city. Her work has won design awards from the Type Director's Club, Art Director's Club, SPD, Print, and Graphics. She has received various celebrated distinctions such as Computer Art's "Top Rising Star in Design" an Art Director's Club "Young Gun", and Print Magazine's "New Visual Artist".

Say What Studio

Say What is a graphic design studio based in Paris, run by Benoit Berger and Nathalie Kapagiannidi, who graduated from the ECV school.

Scandinavian DesignLab

Scandinavian DesignLab is an independent design agency based in Copenhagen, Denmark with representation in Shanghai, China.

Identity is their core business with the vision of building corporate souls, which actually identify and distinguish, and envisioning product brands that connect with the target, build preference and win the battle at the moment of truth.

Among others, their clients are: Nestlé, Unilever, VELUX, Carlsberg, LEGO, VisitDenmark, and Sara Lee. They have won several Danish and international awards, such as The Danish Design Prize, Creative Circle, Clio, EuroBest and the German Design Award.

Seamless Creative

Seamless Creative is a New York City-based, husband-and-wife design team. They specialize in crafting thoughtful brand identities for small and growing businesses.

Shenzhen Huathink Design Co., Ltd.

Shenzhen Huathink Design Co., Ltd. (Liu Yongqing Design Co., Ltd.) is a comprehensive professional design agency specializing in corporate identity design, brand design, commercial real estate, and identity design for city's tourism brand and promotion. It has served dozens of Chinese listed companies and hundreds of large-scale enterprises for their brand image promotion design. Their designs have brought profound influence on the future development of the brands. With 10 years' experience, they have become the first driving force for the rapid development of numerous industry leading brands. They have created many classic branding cases and won numerous awards from design competitions both at home and abroad.

Through the continuous innovation and approaches of systematic integration, they have comprehensive theories, methodology and tools to solve numerous problems during propagating brand image designs, and provide clients with original, forward-looking, systematic and continuous brand image design professional service. It is the belief of Shenzhen Huathink Design Co., Ltd. to cherish every client, solve their problems, create value, and win clients' respect and trust with professional passion and sincere service.

Sergio Mendoza

Born in 1984, Sergio Mendoza is a designer and art director based in Valencia, Spain. After graduated in product design in 2005 he has been tasting experiences (photography, art direction, furniture and teaching) before opening his design office in Valencia in 2010.
Now he combines his time at the studio with the season at Domaine de Boisbuchet (VITRA Design Workshops) and several other collective projects in which he is involved.

Shakira Twigden

Shakira Twigden lives and works in Auckland, New Zealand. She is passionate about all things design, particularly highly illustrative typography.

Simon Guibord

Simon Guibord is a graphic designer based in Gatineau, Québec. He works mostly within the cultural field and specializes in identity design and publishing.

STUDIO NEWWORK

STUDIO NEWWORK is a graphic design studio based in New York. They assemble a team of passionate typographic designers with commitment to search for excellence in design. They design with passion, care, and love. STUDIO NEWWORK publishes a biannual magazine — NEWWORK MAGAZINE.

Studio Botes

Brandt Botes heads up Studio Botes — a boutique design shop in Cape Town, South Africa, that specialises in corporate identities, packaging design and illustration. The studio's work is informed by their sense of humour, love of South Africa, typography and travel.

Synsation

Synsation is specializing in corporate and brand identity design, and with wide-ranging expertise including print design, web development, brand experience and environmental graphic design systems. As design/creative director for international agencies such as VBAT, O&M and M&C Saatchi and also boutique design studios in both Amsterdam and Sydney, Sander Dijkstra — Director Concept & Design now brings his design expertise to Synsation. A diverse portfolio includes work for clients in the travel industry, finance/insurance, hospitality, IT, art/culture, government, automotive, education, FMCG/food, retail, fashion, publishing, telco and business development, with various prestigious international prizes for websites, identities and annual report design.

The Palmetto Press

Owner, Matthew Furr is a recent graduate of the graphic communications program at Clemson University in south carolina. Matthew is often accompanied by close friends Joel Engelberth and Megan Bolick, both Graduate Students at Clemson University. Letterpress is one of this team's many interests and talents and proves to be an ever-expanding hobby. With his precision and love for the art, his clientele continues to grow.

The Office of Gilbert Li

Founded in 2004, The Office of Gilbert Li has gained a reputation for delivering innovative and thoughtful design solutions that successfully engage, entice, and excite clients' audiences. The studio's expertise is in the development of publications, print communications, and promotional materials. From pocket-sized brochures to voluminous coffee-table books, the studio informs all of its work with intelligence, delight, beauty, and craftsmanship.

THERE

THERE is a design agency specialising in creating and transforming brands, developing assets and experiences that enables a better connection with their audience. Services they offer include: brand strategy and positioning, brand identity, brand communications, website and digital media, branded environments, workplace graphics, wayfinding and signage.

THINGSIDID

THINGSIDID, a company would like to record and share art, design & culture. They do not work alone, they have certain cooperation. The outcome is the indicator of what you did and they did. They hope before the end of the world, they could do some nice stuff together.

tind

Manolis Angelakis (b.1981), aka tind, lives and works in Athens, Greece, and studied at Vakalo School of art and design. Manolis uses a plethora of materials, and new and old methods, specifically in the field of silkscreen-printing. Although he was trained as a graphic designer, his knowledge as a printer stems from working as his father's apprentice from a very early age. They currently share their working studio and learn different aspects of the trade from one another. Manolis has also worked as an assisant set designer in several firms such as A touch of Spice and Uranya; under the patronage of the Greek Graphic Designers Association, Manolis has produced and taught master printer/silkscreen workshops, and his work has been recognized by the Greek Graphic Design Awards. He is currently working on his project Error is Superior to Art, which attempts to redirect the public's view and awareness to the value of printing errors and the magnificence intrinsic in them.

TNOP™ Design

TNOP™ Design is a graphic design studio in Thailand, formed by Tnop Wangsillapakun in 2005. They believe that a sharp design is about the critical balance of conceptual work and craftsmanship. Their philosophy is to create a customized design that can represent client's culture through distinct contemporary executions, materials or techniques. The studio has worked with a wide range of clients from a global brand like corbis images to a bespoke suit shop like Alongkorn in Bangkok, as well as collaborating in art and design exhibitions and conducting workshops from time to time. TNOP™ Design's works have won many awards and are featured in many books and renowned magazines around the world.

Toormix

Toormix is a Barcelona-based design studio specialising in branding, art direction, creativity and graphic design set up in 2000. They have worked with clients such as Desigual, Camper, Spain USA Foundation, H10 Hotels, Spanish Ministry of Culture, Barcelona City Council, L'Auditori and chef José Andrés among others. They have recently opened an atelier space for the development of personal projects and research.

Wai Hang Tang

Wai Hang Tang was born in Macao, China. After graduating from Macao Yuehua Middle School, he moved to Hong Kong, China, and then graduated from School of Design, The Hong Kong Polytechnic University. His speciality was graphic design. In the same year, he was employed by the design department of Macao Television as a full-time stage designer. Later, he had served in New York Design in Taiwan, China and then other design studios. In 1997, he established Tong Design Studio, whose works have won numerous awards, and been released on design magazines of China Mainland, Taiwan, Hong Kong, US and Japan etc.

Yanko Djarov

The designer's name is Yanko Djarov and he was born in 1989 in Sofia, Bulgaria. Before university he graduated from a language school and had never been sure if he is to become a part of the digital arts' world. Now he has a bachelor's degree in graphic design and branding and print are his passion.

YesYesYes

YesYesYes is a design studio. After 4 years honing his skills alongside Stefan Sagmeister, Canadian Joe Shouldice opened up this tiny studio in Brooklyn. The work has won most major international design awards, including recognition from the Type Directors Club, Tokyo Type Directors Club, the ADC, AIGA, Graphis, Applied Arts and Coupe Magazine. YesYesYes specializes in branding and all things print.

You & Me

Meeting across a multitude of interactions, they are pulled together to experience the height of all collaborations between You & Me. Who they are? They are a couple. They know what it means to pair two individuals with different potentials to create amazingly groundbreaking work. They are eager to explore a relationship with you as they stretch beyond their own field to discover the best in creativity. Who you are? You are essential in their relationship. You bear a potential that is remarkable and distinct, they salute your individuality. Your uniqueness and their skills work hand-in-hand to spark infinite possibilities. They are awesome together — You & Me.

Yoshimaru Takahashi

Working with his concept that graphic design itself is the culture that speaks of the times, he has been carrying out research into the visual expression of "human warmth" and has been producing works connected with this. Committed to linking design and culture, he participates in exhibition projects, sits on adjudication panels, and gives presentations. He has taken great personal interest in the packaging and publicity of patent medicines, and been involved in exhibitions and publishing on that theme.

Yurko Gutsulyak

Yurko Gutsulyak is professional designer and art director with more than 10 years work experience. In 2005 with his sister Zoryana Gutsulyak he founded graphic design studio — Yurko Gutsulyak (Kiev, Ukraine). Since the studio was established Yurko Gutsulyak was honored with more than 50 awards in design and advertising. Yurko Gutsulyak is actively involved in the development of the local design and advertising market. He is president of the Art Director Club Ukraine (2010 — 2012), moreover he represented Ukraine in the jury of different international advertising and design festivals. In 2010 he was an invited speaker at the European Design Conference (Rotterdam, Netherlands).

About ARTPOWER

PLANNING OF PUBLISHING
Independent plan, solicit contribution, printing, sales of books covering architecture, interior, graphic, landscape and property development.

BOOK DISTRIBUTION
Publishing and acting agency for various art design books. We support in-city call order, door to door service, mail and online order etc.

COPYRIGHT COOPERATION
To further expand international cooperation, enrich publication varieties and meet readers' multi-level needs, we stick to seeking and pioneering spirit all the way and positively seek copyright trade cooperation with excellent publishing organizations both at home and abroad.

PORTFOLIO
We can edit and publish magazine/portfolio for enterprises or design studios according to their needs.

BOOKS OF PROPERTY DEVELOPMENT AND OPERATION
We organize the publication of books about property development, providing models of property project planning and operation management for real estate developer, real estate consulting company, etc.

Introduction OF ACS MAGAZINE
ACS is a professional bimonthly magazine specializing in high-end space design. It is color printing, with 168 pages and the size of 245*325mm. There are six issues which are released in the even months every year. Featured in both Chinese and English, ACS is distributed nationwide and overseas. As the most cutting-edge counseling magazine, ACS provides readers with the latest works of the very best architects and interior designers and leads the new fashion in space design. "Present the best whole-heartedly, with books as media" is always our slogan. ACS will be dedicated to building the bridge between art and design and creating the platform for within-industry communication.

Artpower International Publishing Co., Ltd.

Add: G009, Floor 7th, Yimao Centre, Meiyuan Road, Luohu District, Shenzhen, China
Contact: Ms. Wang
Tel: +86 755 8291 3355
Web: www.artpower.com.cn
E-mail: rainly@artpower.com.cn

QR (Quick Response) Code of ACS Official Wechat Account

Acknowledgements
We would like to thank all the designers and companies who made significant contributions to the compilation of this book. Without them, this project would not have been possible. We would also like to thank many others whose names did not appear on the credits, but made specific input and support for the project from beginning to end.

Future Editions
If you would like to contribute to the next edition of Artpower, please email us your details to: artpower@artpower.com.cn